BY DESIGN

Why there are no locks on the bathroom doors in the Hotel Louis XIV and other object lessons

by Ralph Caplan

St. Martin's Press
New York

Grateful acknowledgment is made
for the permission to reprint
lines from "Heavy Date" from
W. H. Auden: Collected Poems,
edited by Edward Mendelson,
copyright © 1945 by W. H. Auden,
by permission of Alfred A. Knopf, Inc.;
and an excerpt from archy & mehitabel
by Don Marquis, copyright © 1927
by Doubleday & Company, Inc.,
reprinted by permission
of the publisher.

For information, address St. Martin's Press,
175 Fifth Avenue, New York, N.Y. 10010.

Design by Peter Bradford
and Kristen Dietrich

Caplan, Ralph.
By design.

1. Design, Industrial. I. Title.
TS171.4.C35 745.2 82-5656
ISBN: 0-312-11085-5 AACR2

First Edition

10 9 8 7 6 5 4 3 2 1

Contents

Credits

Page 8: Photo by Tony Giammarino; 13: courtesy McDonald's Inc.; 16: courtesy Apple Computer Co.; 28, 83, 153, 154, 171, 174: Maude Dorr; 43: courtesy Time Inc., copyright 1949; 51, 128: Kristen Dietrich; 55: Susan Zeig ; 56, 79, 132, 167: Anne Todd; 59: courtesy Walter Dorwin Teague Associates; 67: courtesy Hyster Co.; 71: Chemex instructional brochure by Peter Schlumbohm; 73: photo courtesy Luxo Lamp Corporation; 76: courtesy Knoll International; 88: courtesy CBS News; 89: courtesy Syntex Dental Products, Inc.; 91: Sandra Williams Associates; 98: "Oil Refinery, Tema, Ghana, 1963," copyright 1971, 1976 The Paul Strand Foundation, as published in *Paul Strand: Sixty Years of Photographs,* Aperture, Millerton, 1976; 115: Mike Schacht; 118: courtesy *Good Housekeeping* (Hearst Magazines); 124: (left) courtesy CBS Inc.; (right) courtesy IBM Corporation; 152: Bruno Massenet; 155, 193: courtesy Herman Miller, Inc.; 159: Henry Wolf; 161: Milton Glaser; 166: courtesy Louis Caplan Grocery Co.; 168: Wayne Thom Associates; 175: courtesy Moshe Safdie and Associates, Inc.; 185: Michael Pateman; 188, 190: courtesy Office of Charles and Ray Eames

Acknowledgments

The preparation of this book was supported by a generous grant from the National Endowment for the Arts. Passages in some chapters appeared first in the magazines *Industrial Design, Progressive Architecture, Design* and *Design Quarterly,* and much of the material in Chapter 4 appeared in a different form in *Chair,* published by Thomas Y. Crowell & Company. Other material first took shape in lectures at various places, particularly UCLA, Carnegie-Mellon University, York University (Toronto), the Cooper-Hewitt Museum, the Smithsonian Institution, the Walker Art Center, the Design Management Institute and the International Design Conference in Aspen.

I am grateful to those organizations, and to Peter Bradford, for designing the book and improving it in the process; Milton Glaser, for designing the jacket; Lonnie Browning and Anne Todd for picture research; Kristen Dietrich for her illustrations; Gayle Sternefeld for finding lost documentation, restoring lost punctuation and generally overseeing the manuscript; Ashton Applewhite of St. Martin's Press for professional and metaprofessional help; Maude Dorr, Michael Pateman and Henry Wolf for special photography; Henry Strub for booking me into the late Hotel Louis XIV as a practical joke and thus leading me to my favorite example of situation design; and Judith Ramquist for her editorial comments and other critical support.

This book is dedicated to the memory of Charles Eames, who did not wait for me to write it.

Introduction

This is a book about design—for people who couldn't care less: the design process has more to do with your life than you may think. It is also a book for people who care too much—who think design is everything: designers won't save the world, but the design process can help make it worth saving. Isn't that enough?

A writer I know decided early in his career that "a subject important enough to warrant a large book should be introduced by a small one." The notion pleased him mightily and he has devoted himself ever since to producing small books on important subjects. The large books he has shrewdly left to others.

This too is a small book on a subject that deserves a large, comprehensive one. It is not a history of design, although there is some history in it; it is not a design critique, although it is critical; it is not a defense of industrial design, although it is sympathetic to design frustrations; and it is not a polemic, although it acknowledges work that is shoddy and mindless.

The point of the book is that design, which is now directed largely to superficial ends, is appropriate to our most significant human activities, and belongs to them. It took me a long time to make the connection, but I have been interested in the design process ever since I heard the following story.

The only theatre in the small midwestern town had no balcony, so there was technically no "nigger heaven." Blacks sat in the last four rows. Not that there were enough blacks to fill four rows of seats, but the house was never full anyway and the management could afford to sacrifice a few rows. They acted as a kind of buffer zone, a token separation to suggest the physical removal that a balcony would have accomplished.

Most blacks in the audience came from the college outside of town, whose students protested against the theatre's policy. They wrote letters reminding the manager that segregation in public spaces was illegal in the state and immoral throughout the entire midwest. They argued that blacks were citizens and that some were

Protest, like the response to protest, can be spontaneous or designed. Here anti-nuclear armament protesters stage a "die-in" at United Nations Plaza. Demonstrations seem chiefly to dramatize a position, but such devices as sit-ins can actually help control a situation.

even war veterans.

None of the letters were answered.

Frustrated, a group of students called upon the manager in person. He was cordial and logical. When they called attention to the law, he called attention to local customs. When they talked to him about morality he advised them that morality could not be legislated. In any case, he explained, the issue was not moral but commercial: if black citizens were permitted to sit next to them, white citizens would not buy tickets. Only when one of the students mentioned a boycott or a lawsuit did the manager show anger.

"Look," he said, "personally I don't care who sits where or does what, but my customers do. Don't you threaten me."

They did not threaten him any further. One evening several months later an unusually large collection of students bought tickets in multiples, although the movie was hardly a major box office attraction. As soon as the doors opened, the students scattered into the theatre in what looked like random clamor but in fact was a well orchestrated seating plan. The last four rows were packed solidly with whites. The rest sat down in racially mixed combinations—two whites and a black, two blacks and a white, never two of the same color together—dispersing themselves rapidly all over the theatre until there was no racially pure area in the entire house. Other customers left in angry protest, but some did not. The projectionist, who happened to be a college student, started promptly, so the movie was on by the time the manager was aware of the problem.

"You people move!" the manager yelled as he ran into the auditorium.

"Move where?" the audience challenged.

"Down in front," was the indignant shout from the four back rows.

The manager tried for awhile to redistribute the audience, but there was too much confusion, too much noise, too much resistance, and anyway there were no empty seats to work with. For him it must have been like trying to work one of those spatial relationship puzzles that challenge you to slide cubes into vacancies that are not yet there. What he faced was *de facto* integration. The only way to resegregate the theatre was to cancel the show. He decided instead to write the night off as a moral defeat (but a financial victory, for the house was packed). Student pranks! He shook his head with almost affectionate intolerance. The next day there would be business as usual, he intended to make sure of that.

The next day, however, bundles of mail poured in from all over the country, and not just from students. These were not letters of protest but letters of praise and congratulations for the manager's heroic reversal of an unfair policy. Telegrams from public figures celebrated his civic leadership. He could not have replied to them all, even if he had known how to clear up the misunderstanding. It was all too complicated. The letters turned out to be just a beginning. The campus newspaper and the national wire services ran identical pieces headlined "Theatre Reverses Segregation Policy."

That theatre never did go back to a policy of segregation, and both the local customers and the manager adjusted, in time, to the new policy. The manager never knew what hit him.

What had hit him was design.

"Design? What has any of that got to do with design? Isn't design the pattern on a carpet or a dinner plate, the shape of a lamp, the styling of cars?

Yes, and the fabric of the carpet; the porosity and temperature resistance of the plate; the lumens of light cast by the lamp, and their intensity; the choices that make the car efficient or inefficient, safe or hazardous, comfortable or uncomfortable.

Design? Isn't design concerned with images–trademarks, corporate logos, record album covers, facades? Isn't it cosmetic and superficial?

Much of it is. And the end product of design very often is surface treatment. So what? We live in a world we look at, so it might as well look good. Most relationships between person and object begin with appearance. Since objects cannot speak for themselves, they need to be made to look like what they are and what they do. That is even true of corporations, which, although they can speak for themselves, are usually incoherent because, in the words of a corporate designer, they are "both impersonal and multipersonal."

But aren't the products of design hair dryers, TV consoles, cereal boxes, curtains, bedspreads, vacuum cleaners, kitchen appliances . . .

Sure. Also chairs and typewriters and computers and office partitions and space capsules and tractors, restaurants and stores and cities, films, books and government legislation and protest strategies.

For design is a process for making things right, for shaping what people need. We all live with designed objects that we love, hate, use, break and don't know how to fix. We need better ones. But we need more than manufactured objects, more than crisp and clever graphics, more than friendly kitchens. We need . . .well, you know what we need. We can get it by design.

The Possibilities of Design

Normal man is designed to be a success and the universe designed to support that success— for . . . man is essential to the success of the universe itself. R. Buckminster Fuller

There is a disease called dermatomiocitis, a collagen deficiency that causes the body's cells to become unstuck. A comparable affliction—the ungluing of the social system—affects the body politic.

Design could help, for design at its best is a process of making things right. That is, the designer, at *his* best, or hers, makes things that work. But things often do not work. And making things right is not just a generative but a corrective process—a way of righting things, of straightening them out.

To be sure, design has not done all that well for us in the past. The acronym *snafu,* devised in the forties as a description of life in the military services, has never had a chance to fall into disuse.

Things are in a mess, and the designer appears to be at least as much the problem as the solution.

Things? What things? Traditionally, designed things are objects—artifacts, tools, hardware, buildings, packages. But things are not what they seem. When we ask someone, "How are things?" we do not expect, or get, an inventory of personal hardware in reply. What we are asking has to do with the situation. Because the question is figurative, it does not seem germane to the design process; but it is.

That is a compelling reason for taking an interest in design, but it is not the first reason. The fact is that we live in a designed world and will never live in any other kind. Even if nostalgia were functional, and we could return to the simpler life we long for and claim to remember, we would have to design it. Equestrian hauling, wind-powered mills, compost-centered communes are now designed ventures, where once they were invented responses to immediate needs.

The book you are reading has been designed from cover to cover, and wrapped in a dust jacket itself designed neither to collect nor inhibit dust but to serve as a poster for bookstore merchandising. The bookstore interior was also designed by a designer. So was every appliance in your house, every vehicle you use to get to work, every machine, every piece of furniture or equipment you use once you get there.

All the hardware of our lives is designed, but what about—to take a horrible but serviceable word of our time—the software? Well, it too is designed. What is a McDonald's hamburger, if not an industrially designed product, the result of market research, facility planning, productivity incentives, demographic studies and product engineering? Each component represents a conscious decision based on testing. The amount of meat is uniform, the quality of the meat is controlled, someone has carefully calibrated the dosage of pickle, onion, and special sauce. Not only is the hamburger a designed product, it is the basic module in a system of designed products, including the assembly line designed to produce it and modify it with options, the extended assembly line designed to get

Hamburgers are designed like other mass-produced objects.

it consumed, and the arches that advertise its availability.

"Software" does not normally mean hamburgers. Ordinarily it means programming. The software for a computer is a program. The software for a TV network is another kind of program. The set for an evening news show is designed (with *hardware:* space-age desks and panels) to position the reporters as serious people at the outer edge of information (software). The set for a morning news show is designed to position the host and hostess as friendly people (semi-software) who may not be expert on world affairs but are cheerful, well informed and compassionate. The set for "The Dick Cavett Show" is designed as a kind of takeoff on living room sets, a parody so verisimilitudinous that it gets confused with the unreal thing it is parodying.

Most of us will never personally require the services of an industrial designer (although the number of people who do is increasing all the time). But although few of us are clients for designers, we get their services whether we commission them or not. In America today, they come with the territory. So we are clients by default and, sometimes, designers by default as well.

It is worth paying attention, then, to what professional designers do to our products and services, and why. Design is not everything, but it somehow gets into almost everything.

Ironically, although design operates in many areas where it is unexpected, it is proclaimed in many areas where it scarcely exists. Because of the glamour image, colors have been transformed by copywriters into "designer colors," just as barbers, after two decades of life as hair stylists, have become "hair designers." Well, why not? Some hair stylists do about as much as some product stylists do, and do it better.

Styling is superficial design, but since the design process operates below the surface, superficial design is a contradiction in terms. It is a contradiction we live with, for design (most of it superficial) is pervasive. How did our lives become shot through with it? Where were we when it happened? Who are the people who do it? Is it good or bad? What can we do about it?

This book tries to address such questions in the conviction that, pervasive as superficial design is, its ubiquity blinds us to its possibilities. The trajectory of design has been moving from the design of objects to the design of the situations in which objects are made and used. And that progression is loaded with promise for us. It is a logical progression. The design ramifications of the telephone have less to do with how the instrument is shaped, or what color it is, than the fact that its very existence makes the user accessible to people he doesn't want to hear from.

Our problems are situational. They always have been, of course, but they have not always looked that way to designers, who really believed that the creation of a better mousetrap would cause the world to beat a path to their door, and that the path led ultimately to the Kingdom of Heaven. We know now that design has consequences. Invent a better mousetrap, and you create the problem of mouse disposal. Incite the world to beat a path to your door, and you create the problems of traffic control, of quality control, of distribution control, of inventory control, of self-control.

These too are design problems, and often more interesting than the product design problems from which, in any case, they can never be divorced. Design is the only way we have of coming to terms with our own technology, and except as we come to terms with it, we live as strangers in a strange land. To the extent that they can make technology understandable, designers work to reduce alienation, for the history of civilization includes a history of manufactured objects. From the crude stone implements of the cavemen to the sophisticated artifacts of modern life runs a procession of increasingly refined technologies and techniques.

Materials technology and manufacturing processes have not evolved evenly. Some of the products we use and depend on most today were hardly able to perform at all in early stages of their evolution. Drip-dry shirts, electronic computers, tape recorders and safety razors have year by year become far better made than they used to be. On the other hand, it is easy for any of us to list products and packages that are not as well made now as they once were—furniture, books, houses, cigars and cigar boxes. Included in such a list would be a lot of things that once were handcrafted but are now mass-produced. There have always been indifferent or poor craftsmen who made inferior objects, but we may be living in the first period in history in which goods that *can* be well made are not. This is not because of indifference or negligence; it reflects an implicit policy based on values and priorities peculiar to our time.

Designers have been blamed for creating forced product obsolescence, and must accept a measure of that blame, although they did not really invent the process; they were merely hired to make it profitable.

When President Nixon exhorted Americans to correct inflation by increasing productivity, *The New York Times* received a bristly letter from a reader who argued it was not productivity that needed to be increased but quality. "It's hard to believe," he wrote,

The entry of the computer into the home is a revolutionary event requiring correspondingly radical design. Because these highly technical devices are used by non-technical people, manufacturers want to make them "user friendly." Designers (including software designers) have to translate that jargon into actuality.

"that amid all the talk of the causes of inflation no one has yet mentioned the effects of the poor quality of merchandise produced in this country. Through built-in obsolescence, the demand for energy and material resources is artificially inflated."

Because we live with goods, how they get designed matters to us all. What should matter even more is that the process by which our goods come into being has some application to social circumstances, and therefore to our lives. It may be hard to see this, because industrial design connotes an arcane, highly technical, narrowly specialized activity. Actually it is, and has to be, the least specialized of professional undertakings, one that laymen invariably find interesting once they see what it involves.

As a layman, I am a sort of case in point. Without any aptitude for, or formal training in, either technology or art, I nevertheless spend a lot of my time working with designers. As a writer I work on projects—publications, exhibitions, films, posters, planning—that usually have some written component. But the writing is subordinate to the conceptual design, which consists of identifying the problem and working out an approach to it. That's what we really do and, for better or for worse, it is done in meetings.

A few years ago I met a woman who described herself as a designer, and I asked her what kind.

"I'm a conceptual designer," she said.

"What does that mean?" I asked.

"I can't draw," she explained.

A *non*-conceptual designer (even if she could draw) would be a contradiction in terms, but most people don't suspect that. I didn't, when the humor magazine I was working for folded, and the only job I could find was with a journal called *Industrial Design*, which I had never heard of and could hardly imagine reading, much less writing for.

As it happened, my first assignment was to cover, "in absentia" through published papers, the International Design Conference in Aspen, Colorado. I took the papers home and read them. They dealt with design in so broad a sense you would have had to be dead not to be interested in the subjects. The papers were a positive critique of civilization. Was *this* what designers were engaged in?

Unfortunately it wasn't, and still isn't. But behind that sobering fact lay an inebriating possibility: it *could* be what designers are engaged in; and if it were, we would all profit. Before that can happen we need to be saved from Design, and designers need to be saved from us. By "Design" I mean what used to be, and too often still is, called "Good Design"—the expressions of taste manufactured by tastemakers, the collections of certifiably acceptable objects (some of them marvelous indeed) that are the material counterparts of the reading lists approved by the Great Books clubs.

By "us" I mean all who, as consumers or clients, conspire with designers to misuse their uneven but powerful array of talents.

Although I knew hardly anything about industrial design when I stumbled into it, this was not the handicap it might seem. As it turned out, designers did not know much more than I did about the profession they practiced. Moreover, one of the things they did know about it was that their very ignorance of the details of a client's business was sometimes an advantage. So was my ignorance of design. (While technical ignorance can be an important attribute of design capability, this is not to argue for a know-nothing school of design. Ignorance is useful only at the beginning of a project. If it lingers to become what Roman Catholics call "invincible ignorance," the consequences can be tragic.)

I learned more about design after I stopped writing about designers and began working with them. There is very little information available about how design gets done, about what really happened on particular projects. Design projects are almost never reported in the kind of detail that, say, *The Wall Street Journal* routinely provides in reporting a corporate takeover; and —particularly in the United States—there has rarely ever been serious design criticism. That is changing. I once complained in a speech:

> One kind of restraint from which designers are curiously free is the force of critical scrutiny. There is almost no public design criticism at all, except in architecture. Every novelist, every playwright, every composer and musician, every painter, actor, prize-fighter, filmmaker, every chef, offers his work with the knowledge that it may be praised, blamed or analyzed. But when a manufacturer throws yet another refrigerator into the stream of economic life, he is assured of exemption from critical comment. Any number of journals may report the product's availability—often in the same terms in which the manufacturer reports it—but no one reviews it.

As it happened, Dexter Masters, then editor of *Consumer Reports,* was in the audience. "We're way ahead of you," he said.

They were. Years before, Masters had got the architect Eliot Noyes to do precisely what I was talking about. Noyes was a first-rate design critic, and performed as one for his clients. When he stopped doing the *Consumer Reports* articles, Masters was unable to find another designer to replace him because the professional industrial design societies said it was unethical for members to criticize each other's work.

Perhaps for the same reason that evangelism is more fun than ministering to a congregation, I have never particularly wanted myself to write the kind of thing I was urging on everybody else. However, I did write a piece for *Consumer Reports* on automobiles. The vigorous reader response, some of it hostile, confirmed my notion that there might be high interest in critical writing about industrial design. Since then, supported by such social forces as Ralph Nader, the women's movement, inflation, and student disenchantment with material values, product criticism has come into its own. Television has carried it much further than newspapers or magazines, and is potentially a better medium for demonstrating what's right and wrong about the things we live with.

Television stations throughout the country now have consumer reporters, many of whom enjoy the right that movie reviewers have always had, the right to be candid even when they are reviewing advertisers. In Boston, WBZ's Sharon King pioneered the direct challenge to advertised superiority. In the spirit of science and entertainment, she began trying to duplicate the demonstrations on TV commercials. Her experiments did not necessarily get the same results that the advertisers did: the dogs failed to go for Alpo, sheets of Bounty crumpled under the weight of a cup of coffee. ABC's John Stossel takes periodic hard looks at products and services, and NBC's Betty Furness, who became famous proclaiming that "you can be sure if it's Westinghouse," has made a contradictory second career of demonstrating that in the world of consumer products, you can never be sure.

Serious criticism is predicated on the assumption that the activity under discussion is itself serious. Industrial design (with good reason) rarely looks serious. As a matter of fact, it rarely looks any way at all. Considering that designs are everywhere, what designers do is neither highly visible nor widely understood. We are all aware of the results but not of the activity.

Several years ago I wrote the text for a traveling United States Information Agency design exhibition directed to communist countries.

"Please keep the text simple," the project chief admonished. "Bear in mind that this is going to be read by Polish peasants who don't know anything about American industry or about design."

I did bear it in mind. So far as I know, the text did not confuse any peasants. It did, however, attract the editors of an American business magazine, who translated it back into English and published it under the title "What Executives Should Know About Industrial Designers." It has since been reprinted under that title a number of times by business publications here and in Germany and Japan. Another version of the same article was published in an engineering magazine under the title "What Engineers Should Know About Industrial Designers." The moral is clear: design for American industry is so little recognized that information addressed to Polish laymen turns out to be news to management and engineers.

Small wonder; for we are *all* laymen. Perhaps nothing about the industrial designer is more important than the fact that he or she is a generalist. It is easy for everyone, including designers, to overlook this but it is at the heart of design as a process of potentially great social use.

Design is not such a process now. Even more than advertising, industrial design is a profession full of sensitive, intelligent, highly trained men and women bringing their sensitivity and talents to bear on trivial problems. Furthermore, unlike advertising people, industrial designers are trained or experienced in a process appropriate to significant problems. In wasting them—with their collusion—we waste not only our resources, but ourselves.

The Professional Emergence

Here is Edward Bear, coming downstairs now, bump, bump, bump, on the back of his head, behind Christopher Robin. It is, as far as he knows, the only way of coming downstairs, but sometimes he feels that there really is another way, if only he could stop bumping for a moment and think of it. A. A. Milne

The industrial designer's job was to introduce art into industries hitherto artless. Fortune, 1934

Most of the material in this book is American. Industrial design is not better done in the United States than elsewhere, but it is more firmly established. So are industrial designers, although they, like their discipline, have been regularly undergoing change for the past few decades. Because of the particular direction that industrial design has taken here, American designers have both more freedom and more restraints than their European or Asian counterparts.

An English journalist claims that "the degree of influence which American designers have over the big corporations which employ them astounds the British." American designers are astounded to learn of their influence, for they know that the strength of design influence in American industry is illusory. Nevertheless, American designers have made themselves more conspicuous in industry than their European counterparts, and European designers have performed—or at least written and spoken—in more academic terms. (Even when a word like *educating* comes from the lips of American designers, it is likely to mean persuasion or selling, as in "we've got an educating job to do.")

The distinction has not always been viewed as academic. As the Common Market prospered in Europe, American businessmen worried. One of the most important things they had to worry about, according to Walter Hoving, who as president of Tiffany undertook to worry about design from time to time, was "our comparative weakness in the field of industrial design." Hoving understood industrial design to be concerned almost purely with esthetics, a word he used about as often as teenagers say *like.* After railing against the esthetic inferiority of American products, and the American client's indifference to and ignorance of esthetics, Hoving came up with a solution: "the creation of a new profession which will function midway between top management and the design staff."

What this comes down to is the invention of a middleman to stand between other middlemen. The consumer, about whose arrogance in the face of ignorance Hoving also expressed dark misgivings, is thus even further removed from the people who fashion the products he is going to buy. Apparently distrusting designers almost as much as he distrusted consumers and clients, Hoving proclaimed: "I do not believe the solution is just to turn everything over to the designers and give them a free hand. They're generally proficient in designing only the products they have been trained to work on."

That is a remarkably lucid expression of one of the most common misconceptions of design, for no industrial designer worth his salt, or our attention, has been trained to work exclusively on *any* particular product, unless by accident. What he has been trained to do is practice a process called design, a process that includes esthetic choices, but does not consist only of them. Industrial design is relatively uncomplicated, even though it is hard to describe. There are myths about industrial design that deserve examination. As background, however, it may be useful to look at some of the industrial designer's assumptions about *us.*

The consumer is a woman. There is less sexism in design than there used to be, but there is still enough to encourage patronizing the buyer. It is not surprising that designers of consumer goods think of the consumer as "she." Women do buy most packaged goods and household appliances. But thinking of *her* liberates the designer (usually *him*) from the necessity of identifying with her. The closest he comes is identifying his wife with her (which may be what's wrong with his marriage). Consequently designers are often unable to see clearly the people they are presumably designing for.

This is as good a place as any to point out that, just as the consumer may not be a woman, the designer may not be a man, particularly in the fields of interior design and graphics, but also in architecture and industrial product design.

Consumers have poor taste, but can gradually be led to better taste by designers. Consumers have no taste—that is, no fixed collective taste. They like some things and don't like others, and their likes and dislikes are subject to a wide variety of influences: age, education, advertising, peer pressure, ethnic background. But the greatest influence is availability. It is a matter of public record that no consumer has ever chosen a second-rate product that didn't exist. If it exists, it was designed by a second-rate designer. Some designers, to be sure, argue that they are merely satisfying the vulgarity of the market, but satisfying someone else's perceived vulgarity is a vulgar act.

Great amounts of money and special facilities must be spent to find out what human beings are like and want. Genuine design research is necessary and rare, but a lot of spurious research is directed to creating mock mysteries instead of dispelling real ones. Since designers are, as a rule, people, and since individual designers are, as a rule, persons, one needn't look quite so far to find out what people want as marketing folk suppose. The difficulty lies not in learning what is wanted, but in being able to provide it.

Yet conventional design wisdom holds just the opposite: if only we knew what they wanted, we could do it. Not so. As Old Lodge Skins in Thomas Berger's *Little Big Man* says, "Sometimes the magic works, sometimes it doesn't." Researchers keep insisting that "we need to learn more about people," as if people were some exotic species. But as the black comedian Godfrey Cambridge told his mother when she warned that "they" were moving into the Cambridges' neighborhood, "Momma, *they* is us!"

There is no disputing that consumer preferences are sometimes surprising in ways that run so curiously counter to common sense they could only have been revealed through formal research. But much of the market research that finds its way into design is what I call "justification research," aimed not at supplying genuine information or insight but at supplying a rationalization for doing something fairly obvious.

Let me give you one of my favorite examples. One of the principals of a large package design firm invited me to sit in on a client presentation. I arrived a few minutes late, and was ushered into a dark screening room, where the presentation was already under way. On the screen were projected various designs that had been tried out on test audiences and rejected. The man doing the presentation had a delivery more polished than I had expected of a package designer; he sounded like an actor playing a designer making a presentation. With the aid of slides, he led us through various trials undertaken over a period of several months in an attempt to establish what consumers wanted in a salad dressing. The researchers had at last succeeded in isolating the qualities important to housewives. They were "goodness" and "creaminess." As the package designer announced this, the words themselves appeared on the screen to reinforce the idea: GOODNESS CREAMINESS. The presentation went on to show how that revelation had led designers to create a label emphasizing goodness and creaminess. At this point there was an interruption.

"Harry," said a voice from the floor, "Don't you think we ought to play up the idea of competitive packaging more?"

"Maybe, Chet," the narrator said, "but I thought that would come across better at the end. It's near the end of the script."

Harry? Chet? Script? Suddenly I knew what I would have known from the beginning, had I arrived on time. I was not sitting in on a client presentation. I was attending a rehearsal! The speaker, I learned later, was not a designer but a man who had been hired to make pitches to clients.

After the run-through I was asked by the firm's partners what I thought of the presentation. "Interesting," I said. "But the design looks to me pretty much like other salad dressing labels."

"Yes, we didn't want to move it too far forward. Our research indicated conservatism."

"About that research," I said. "Honestly, didn't you know to begin with that people wanted goodness and creaminess in a salad dressing?"

"How could we have?" he asked. "No one knew."

"Well, did you think people wanted badness and thinness?"

The other partner spoke up. "I know," he nodded. "It looks obvious now, after the fact. But in today's marketplace we can't afford to guess. There's no room for intuition. We have to be sure."

Yet designers have traditionally incorporated their intuitive understanding of people into designs. That is what they are for. What affects us so strongly when we see an MG or a beautifully balanced knife is that someone has pleased us by making what we wanted and never knew we wanted. This ability to transform empathy and understanding into serviceable form makes designers potentially far more useful citizens than they are at present. Designers say a lot, if not always clearly, about what they think we are like. What do laymen think design and designers are like? There are some clues in the questions that get asked.

Henry Dreyfuss was sometimes called "the dean of American industrial designers." Walter Dorwin Teague, a contemporary of Dreyfuss, was also called that. I don't know why industrial design should have two deans—I don't even know why it should have one—but of the two, Dreyfuss was more deanlike: mellow, eminently successful, warm and accessible to colleagues, although less intellectual in style than Teague. Maybe Teague was Dean of Academic Affairs and Dreyfuss was Dean of Students. In any case, Dreyfuss, who was available to the curious for most of his life, reported in all seriousness that one of the questions most commonly asked when people learned he was an industrial designer was, "Why are barns painted red?"

This stunned me when I heard it. I can imagine someone asking the question once, although even that seems unlikely, but I never have been able to believe in the commonness of the inquiry. No one has ever asked me why barns are painted red, and a very good thing too, since I don't know (neither did Henry). Here are some questions I *have* been asked.

Q. What is design?

A. That's a terrible question, but I suppose an unavoidable one. This book, which is passionately non-definitive, is one answer. In the meantime, if pressed, I would describe design as the artful arrangement of materials or circumstances into a planned form. If an arrangement is unplanned, it is not design, for design implies intention and purpose. If an arrangement is not artful, it is not design, for design employs art skills (however minor) in planning and in executing a plan.

Q. What is "good design?"

A. A red herring. "Good" design as a partisan cause is no different from good sex or good food. No one wants bad *anything*. Design solutions ought to solve problems appropriately as well as effectively; *appropriate design* is a better objective than good design, because there are meaningful criteria for determining it, and because it includes the reality of change and circumstance. (Besides, market research indicates that appropriateness and creaminess are the qualities most desired in a designed product.)

Q. What is industrial design?

A. A misnomer. Originally, it meant relating people to things through the shaping of goods for mass production (but exhibits and feasibility studies aren't mass-produced). All right, then, industrial design is design for industry (but also for government and for nonindustrial organizations). You see the problem. Industrial designers do more than the definition implies. They also too often do less. Industrial design is what industrial designers do, but they are not the only ones who do it. In an industrial civilization, perhaps all design is industrial design.

Q. Are industrial designers homosexual?

A. This was the first question asked me at the end of a talk about industrial design, and it really threw me. The questioner was a woman whose refrigerator bore an escutcheon plate that collected dirt and resisted cleaning, but served no other purposes. She had noticed that her other appliances had also been gratuitously made hard to clean and to use, and she concluded they could only have been designed by people who hated women. To her that meant male homosexuals.

Were the products she complained of really designed by people who hated women? The odds are that they were designed by people who did not think seriously about how they would be used by anyone. Yet the design process carries within it the germ of this understanding. It isn't necessary to do research to find out that people want a clean refrigerator. But it is necessary to consider the characteristics of the user, and this does call for direct, nonscholarly research. The questioner may have been naive about homosexuality but she was right about her kitchen.

In *The Hidden Dimension,* social psychologist Edward T. Hall writes:

> My wife, who has struggled for years with kitchens of all types, comments on male design in this way: "If any of

Men's feet and women's feet are similarly built but differently shod.

the men who designed this kitchen had ever worked in it, they wouldn't have done it in this way." The lack of congruence between the design elements, female stature and body build (women are not usually tall enough to reach things), and the activities to be performed, while not obvious at first, is often beyond belief. The size, the shape, the arrangement, and the placing in the house all communicate to the woman of the house how much or how little the architect and the designer knew about fixed-feature details.

Now that the "woman of the house" is not the only household member to occupy the kitchen, have fixed-feature details improved? Not necessarily. Designs that don't work are not necessarily attributable to the designer's ignorance or indifference. Sometimes they can be laid to the triumph of fashion over common sense.

Q. Why don't they make women's shoes that are comfortable?

A. "They" do, more or less. But many women would rather endure the pain than wear the shoes. If a visitor from outer space were shown a man's brogue and a woman's high-heeled, pointy-toed capsule he would assume that the two shoes were made for feet radically different in shape. There are not, however, significant anatomical differences between men's feet and women's feet. A man's shoe is designed at least roughly for the shape of the foot it encases, and a woman's shoe is designed with utter disregard for that shape, conforming instead to the shape of feet that exist in the fantasies of a fashion designer. Bad design hurts.

One of the curious features of the designer's life is what he designs for himself. For all the talk about conversation groups, furniture is arranged in ways that make sitting, eating and talking difficult. We have of course all heard that electricians live in the dark, that carpenters don't have enough closet space, that shoemaker's children are unshod, but this is something else—a self-conscious "statement." Landscape architect Paul Friedberg recalls designing children's rooms that visiting designers and editors admired, but that were inconvenient for his children, and designing a country house that had no back door until his wife made him break one through the wall. For the sake of design integrity a professional designer will put up with vast amounts of discomfort, and resent any consumer unwillingness to make the same sacrifice. When someone criticized the inconveniently low ceilings in Ford cars, styling director George Walker was quoted as saying he found it hard to believe anyone would mind getting his hat knocked off for the sake of a clean line.

A young designer gave me his business card, which I started to insert in my wallet, only to discover that it did not fit there: it was too big. Yet the type size was too small to be read without difficulty. A card too big for a standard wallet ought at least to be legible. I started to put it into my shirt pocket, but since the shirt had been designed by a designer, there was no pocket.

Once their taste has been pounded into shape by tastemakers, consumers will put up with the same nonsense designers do, as Randall Jarrell observed:

> The public that lives in the houses our architects design . . . has triumphed over inherited prejudice to an astonishing degree. You can put a spherical plastic gas tower on aluminum stilts, divide it into rooms, and quite a few people will be willing to crawl along saying, "Is this the floor? Is this the wall?"—to make a down payment, and to call it home.

Q. Is industrial design a kind of engineering?
A. No, but it can't get very far without engineering. See Chapter 5.
Q. Is design art?
A. Yes. I mean no. See Chapter 6.

Historically, industrial design shapes objects that are manufactured by machine rather than crafted by hand. With the movement of machine-made goods as an economic mainspring, and with the relationship between maker and buyer growing increasingly complex and increasingly remote, the designer's importance looms large. Industrial design projects range from clothespins to railroad freight car interiors, and also include such apparently peripheral activities as corporate communications, feasibility studies, product planning, human engineering and space planning.

Because of this range the industrial designer has often been called (by designers or their press agents) a "Renaissance man." The term is most likely to be used this way by people who don't know what, or when, the Renaissance was. Renaissance Man, the real one, had more going for him than mere versatility; and the fact that a designer works on a logo for a clothing manufacturer one day and a logo for a television network the next, does not make him Francis Bacon.

American industrial designers do perform an astonishing breadth of activity. To some extent this reflects a corresponding acceptance of design breadth, but it also reflects a desperate willingness to get the designer's foot into any door that will open. Designers, starting out to perform a service called design, discovered that no one understood it, no one wanted it, and there wasn't much money or prestige in performing it in any case. So—just like the bookstores that sell gifts and games and greeting cards because they found no market for books—they began doing other things, or claiming they could. "Once you get their confidence," a Chicago designer says, "clients ask for advice on anything—like, can you recommend a dentist? You find yourself saying, 'Well, I'm a pretty good dentist. Why go outside? . . .' "

Dentistry may be far afield, but it is reasonable for industrial design, like law, to shape itself to the needs of clients; also to shape itself along the lines of the designer's interests and abilities. In a lovely remodeled church in Santa Barbara, California, the firm of McFarland, Bartlett, Tanner is in the business of product development.

"We *do* industrial design," Don McFarland explains. "In fact we won't do anything that doesn't have some industrial design content. And for some clients we do straight-out industrial design and no engineering. But although we sell industrial design services separately we do not normally sell engineering services separately, because if you turn yourself into an engineering operation you do nothing but solve heat problems or mechanical problems or electrical problems, and we're end-product oriented."

As a design executive at General Electric, McFarland had been in charge of advanced engineering and industrial design and the present partnership combines those disciplines. "The only way you can do everything from concept to product development is to put the right people together," he says—"designers, engineers, modelmakers, machinists. Our clients don't need this kind of creative team in house, although many of them do need in-house industrial designers."

While the roots of industrial design are the roots of the industrial revolution itself, the most conspicuous flowering began in Germany in the 1920s with the formation of the Bauhaus, which was a school of design in both the formal and figurative senses. Led by such architect/designers as Walter Gropius, Mies van der Rohe, Joseph Itten, Joseph Albers, and László Moholy-Nagy, the Bauhaus came to have the kind of effect on design that Freud had on psychiatry. That is, it gave technology a coherent—if humorless—philosophy. As with Freudian analysis, it did so at a price: the acceptance of a dogma from which architects and museums still find it hard to escape. (A later penalty was exposure to Tom Wolfe's caustic wit in *From Bauhaus to Our House.*)

The most popular expression of that dogma was the dictum "form follows function." That succinct phrase, from the Bauhaus leaders in Europe and from Americans like the sculptor Horatio Greenough and the architect Louis Sullivan, was for a while almost the only statement of design philosophy heard in the United States. As the machine displaced crafts and designers sought desperately to determine what machine-made objects ought to look like, the purpose of these objects had sometimes taken second place. In their passionate effort to understand and come to terms with the machine, Bauhaus leaders insisted that the form of an object must be conceived in terms of what that object was intended to do. Design integrity became a high moral imperative, and a dishonest chest of drawers was equated with personal dishonesty and might have been assigned a scarlet letter if graphic designers were not so busy debating the evils of the serif.

The notion that form follows function needs to be qualified. It seems initially to have been a corrective admonition: designers had better attend first to how an object works, *then* to its shape and appearance. But "form follows function" was quickly interpreted to mean that if an object were made to function well, it would *as a matter of course* be appropriate and pleasing in appearance. This is demonstrably untrue. A sampling of any year's patent office drawings will reveal a wide variety of products in which function has been followed by forms truly dreadful to contemplate. So will an inspection of your own kitchen or garage.

While regard for function does not guarantee good design, it is obviously true that to ignore function is to guarantee bad, although not necessarily ugly, design. The sad truth is that a great many functionally ineffective designs have won "good design" awards.

That form follows function is, as critic Edgar Kaufmann, Jr. has pointed out, merely a statement of sequence: form *follows* function. What function dictates is not form, but a set of boundary conditions. Within the constraints imposed by these conditions a design may take any of a variety of satisfactory forms, depending on the technology and materials available and—above all—on the designer's talent. "Form follows function" is still a useful way of describing order of concern—an order that in an ideal society would be too obvious to mention, but that in ours needs constant reiteration.

The American industrial designer, like any other culture hero, real or synthetic, rose in context. To a humbling extent, design for industry was merely The Kind of Thing People Were Doing, as reflected in this passage from *The Group,* Mary McCarthy's novel about Vassar girls in the thirties:

> Connie Storey's fiance, who was going into journalism, was working as an office boy at *Fortune,* and her family, instead of having conniptions, was taking it very calmly and sending her to cooking school. And lots of graduate architects, instead of joining a firm and building rich men's houses, had gone right into the factories to study industrial design. Look at Russel Wright, whom everybody thought quite the thing now; he was using industrial materials, like the wonderful new spun aluminum, to make all sorts of useful objects like cheese trays and water carafes. Kay's first wedding present, which she had picked

> out herself, was a Russel Wright cocktail shaker in the shape of a skyscraper and made out of oak ply and aluminum with a tray and twelve little round cups to match—light as a feather and nontarnishable, of course.

Vassar has changed since then, and so has American design. If Russel Wright in 1933 was bending industrial energies to the problem of cocktail party accessories, in 1956 he was dispatched to Southeast Asia by the Mutual Security Program in an effort to determine, and help develop, the economic potential of emerging nations.

American industrial design was first officially recognized in 1925 when the United States sent a commission to the Paris International Exhibition. The commission's purpose was urgent: to alert American industry to the necessity of meeting foreign design competition and to acquaint American industrialists with the problems peculiar to techniques of mass production. What gave American design its real chance in life, however, was a phenomenon not on display in Paris: the Depression.

This ought not to have been surprising. In 1915 Henry van de Velde, director of the School of Arts and Crafts in Weimar, had written to the architect Walter Gropius: "As long as things are going well for the industrialist or the craftsman—that is, as long as he finds a market for his goods—no matter what sort of quality they may be, he will not consult the artist. But when he does, when things are going badly for him, he seeks him as he would seek the devil to sell his soul, determined to do anything to fill his threatened purse. And if the artist fails to help the manufacturer sell a hundred of each of the new models at the next Leipsig Fair, then the time is up for the devil or the artist!"

That is why artists were sought out during the economic crises of the early 1930s in the United States. Harvard economist M. W. Sprague wrote that "failure of industries to adopt policies designed to open up additional demands for industrial products is, in my judgment, the chief cause of the persistence of the Depression." But how to create such demands? No one knew. No one even knew how to find out. Desperate straits called for desperate measures: industrial designers were sought to stimulate sales by making products attractive. It seemed to work—at least the Depression finally wound down—and industrial designers got a good deal of the credit (probably far more than they deserved) for the economic revival. Industrial design's presumed value to the industry was based on the premise that "styling" sells merchandise. To a great extent it does. As does advertising. As do "after Christmas" sales.

This, to be sure, was not the most promising of professional origins. The marriage (some think unconsummated) of industry and industrial design was a marriage of convenience that often turned into a marriage of inconvenience. Designers were like stereotypically glamorous women who turned out to be good cooks and responsible homemakers as well. Moreover, the most interesting of them discovered liberation and insisted on having ideas, launching careers, initiating projects and generally raising hell. Such behavior irritated husbands and industrialists alike. Designers have at times yearned to be the conscience of industry, to the consternation of clients who could not recall having retained any consciences. Clients were at times astonished by the experience of hiring a designer to do something simple like change the housing on a humidifier and finding that what they had hired instead was a busybody who wanted to work on new ways to facilitate breathing.

Nobody-asked-me-but is a common design posture. In 1959 *Fortune* magazine published an article called "Odd Business, This Industrial Design." The oddest aspect of the business, according to the author, was the designer's propensity for expanding a simple request for the face-lifting of a product into the redesign of the corporation that manufactured it. The article begins with a longish anecdote about the president of a tool manufacturing firm who called in a designer to help meet the competition of a better designed paint sprayer. The designer responded by undertaking an intensive research program, resulting in a detailed analysis of his client's sprayer and of competitive sprayers, and a number of recommendations. One recommendation was that a product planning committee be set up so that the designer and the president could discuss future projects. "I sometimes get the feeling that he isn't satisfied to be a designer," the president said. "He wants to be my right hand and maybe even me." According to *Fortune,* "the incident is fairly typical."

The historical background of industrial design is relevant because the schizophrenic nature of the profession today reflects its split beginning. Philosophically industrial design began as a profession of lofty motives; but in practice it was almost always becoming something else because something else was wanted. The Bauhaus roots make design as a profession far different from advertising, with which it is sometimes compared. (Such comparisons may be favorable or unfavorable. The New York graphic design firm of Lippincott & Margulies candidly based its organizational structure on that of an advertising agency, with account executives to "service" clients, plus a legion of motivational specialists and marketing analysts of the kind that advertising agencies use. The comparison is made less charitably by Victor Papanek in *Design for the Real World:* "Advertising design . . . is probably the phoniest field in existence today. Industrial design, by concocting the tawdry idiocies hawked by advertisers, comes a close second.")

Advertising began simply enough as a means of announcing what was available in the marketplace, then developed into a means of inducing people to buy one product rather than another. Later refinements saw it mature into a means of inducing people to want things they didn't need, to buy things they didn't want, and to covet things they were not yet ready to buy. As the techniques became more obvious and the influence of advertising became more pervasive, a certain public resentment set in. In addition, many people in the industry itself felt uneasy in a business based largely on appealing to greed and stimulating fears. Both for public relations purposes and from a sense of their own dignity, advertisers undertook to do good works. They still do. And advertisements for the Muscular Dystrophy Association or the Red Cross get duly credited in service-to-mankind ledgers out of decency and guilt. Individual advertisers may have felt they were prostituting their talents, but the industry as a whole was unembarrassed, having had not very far to fall.

Industrial design *had*—at least in theory—although that was not always apparent. The thought of styled obsolescence would have been abhorrent to William Morris and the Bauhaus leaders who followed him, and is abhorrent to a great many designers today. But even the most sophisticated businessmen were willing to believe in styling as a solution to their problems. And some designers did too. In 1934 *Fortune* ran a photograph of a wild jumble of hardware that looked like the interior of Fibber McGee's closet, but was in fact the waiting room of industrial designer Lurelle Guild. The caption, headed NEW CANDIDATES FOR FACE-LIFTING PILE UP IN MR. GUILD'S WAITING ROOM, reads: "A few of the thousand products redesigned by Lurelle Guild in a year are here seen ready to take the art cure. Soon one of them will be pulled, Guild will glance at it from his drawing board, make a dozen sketches, circle the best four, make finished drawings of these, send them to the client, who will select one, and put it into production."

Did it ever *really* happen that way? Almost surely not. Although the early designers were concerned with style, even from the first they knew better than to stop with styling. It is a testimony to their astuteness that, although their method was largely intuitive, the first designers insisted they could do no effective design without a solid understanding of the client's problem and the information that bore on it.

At a time when manufacturers wanted nothing more than a "face-lifting," an "art cure," a prettification of what their engineers had devised, the two most stubborn resisters came from the presumably capricious world of the theatre. Norman Bel Geddes and Henry Dreyfuss had both begun their careers as theatrical set designers. As early as 1929, when the Bell Telephone Laboratories offered Dreyfuss a $1,000 fee to make some sketches of what telephones should look like, the designer refused on the grounds that the fee was inadequate to provide a serious study of how telephones were made and used. Without such a study, Dreyfuss argued, sketches would be irresponsible.

At about the same time, the flamboyant Bel Geddes was approached by the president of the Standard Gas Equipment Corporation, who complained that his line of stoves was, for the first time, being hard hit by competition. He wanted Bel Geddes to design a new stove and to provide drawings for it within a week or two. Bel Geddes replied that if he could find a way to genuinely improve upon the stove, the job would take a year. When he got the commission, he installed an engineer from his own staff to work with company engineers; a team of Bel Geddes designers made studies of all competitive stove models and another team interviewed dealers and housewives. Such thoroughness enabled him to reduce hundreds of stove parts to 16 basic modules, a change the corporation resisted at first, although production costs were cut drastically. Sales for the new stove rose, then doubled.

Almost the same story, with different characters, was told by Walter Dorwin Teague. By the mid-twenties Teague was a prosperous, established graphic designer, with a particular reputation for fancy book design, but "dissatisfied and bored with the work I was doing." He designed a few packages. He sketched a line of automobiles. A piano manufacturer asked him to design some grand pianos. While Teague enjoyed the projects, he felt that none of them had worked out very well, except for the grand pianos, which he had built under his own direction.

In 1927 Henry Ford announced that the Model T would be discontinued and the plants closed for a year and a half while a new car was designed and production tooling prepared. That started Teague wondering how many other companies were, in effect, making Model T's. At about this time the conservative Eastman Kodak Company approached Teague about the redesign of two cameras. Remembering what he had learned from designing the grand pianos, Teague told Eastman that "I knew next to nothing about cameras, but I could only undertake the assignment if I could do the work in their own factory, working with their engineers; and I would need a year's contract for one week a month and I set a fee high enough to make sure that my opinions would be respected." The company accepted, and for decades afterward the Teague office designed Eastman Kodak equipment.

Unusual but not unique. In 1940 the trade magazine *Electrical*

Manufacturing reported the design of a new line of vacuum cleaners for Montgomery Ward. The designers began with a study to secure data on the features of all competitive products. The design goals growing out of that study could have been written by Ralph Nader: "increased cleaning efficiency, better appearance, low cost, reduced operating noise, convenience in operation, lighter weight and maximum accessibility for servicing."

What those early designers seemed to be saying was that, in order to do a good job, they had to know what they were doing. It hardly seems to be a profound point. The reason it was so difficult to understand and to accept, has to do with a basic—and not entirely unjustified—distrust of the idea of the "artist in industry." According to August Heckscher, "The whole difference between the modern age and the older ages of the world can be summed up . . . in this way: the modern age designs things; the older ages made them. If you really make a thing, it doesn't need to be designed. It has an ultimate and fitting shape, its purpose and appearance are inextricably related."

I should prefer to say that people used to design and make things; now some people design them and other people make them. When the craftsman was maker, there was no need for an artist in industry. The need developed as industry developed: the industrial revolution, when it substituted factory for craftsman, separated the designer from the process of making products. What had once been a one-person and, therefore, a naturally integrated affair, became too complex for one person to control. The temptation, the natural impulse, was to put the designer back, but at the tail end of the production process. So it was perfectly natural that the earliest design for industry was "applied art"—art that was *literally* applied: simply added to products after the fact.

The unsatisfactory result was a spate of machine-made products that imitated their hand-made predecessors. While this was understandable in the nineteenth century, which had not yet begun to learn how to exploit the machine sensitively, it is less easily defended in the 1980s. It persists today in the use of plastics to imitate leather and wood, and in the machine manufacture of furniture copied from old handcrafted products. In late nineteenth-century England, the Arts and Crafts movement, led by William Morris, revolted against the derivative cheapness and ugliness of machine-made products and attempted to resurrect the ideals of good craftsmanship by resurrecting the craftsman himself. While this was impossible on the scale desired—there was to be no turning the industrial clock back—the movement did focus attention upon the search for a solution in the nature of the machine itself: the Arts and Crafts movement pointed directly to the Bauhaus.

But not many artist's studios had machines (as compared with today, when an artist's loft may have as much machinery in it as the job shop that occupied the loft before the artist moved in). If designers were to find integrity in machinery, they needed high-level access to the manufacturing process. They needed power. Who could give it to them? Only industry. So recognition by industry became imperative as a professional concern of designers.

One respect in which American industrial design has won industry recognition is the steadily increasing use of designers for capital goods and for nonconsumer products generally. The design of machine tools, laboratory instruments, and electronic data processing equipment has encouraged improved design for the consumer markets. Indeed, as the "high-tech" trend indicated, consumers have uncovered one of the secrets designers have long cherished: materials and products made for industrial use are better, cheaper, and handsomer than those made to sell in stores. Lights, shelving and cleaning equipment move from warehouse to our house with surprisingly few problems of adjustment.

Tools demand serious design. Industrial equipment is more expensive to make and buy than consumer equipment, is more hazardous to operate, and far more depends on its effective operation. Badly placed controls on a stereo tuner are a nuisance. Badly placed controls on a turret lathe can be a disaster, resulting in accidents, lowered productivity and defective products.

Design is important in industrial equipment for an additional reason: people buy it, in part, for the same sort of emotional reasons they buy anything else. The stereotype sees the consumer as a frivolous person subject to the appeals of style and status, as compared to the rational professional or industrial buyer. Anyone who truly thinks this has never seen an executive choose an attaché case or a desk, has never seen a farmer buy a tractor, has never met a colonel in charge of purchasing military equipment. They, like anyone else, make choices partly on the basis of how a product feels and looks. Even a purchasing agent specifying a commodity brings *some* personal choice to the transaction. These are not irrational decisions; they are merely the irrational component of rational decisions. Once it has been demonstrated that a copier or a tractor will do its job, once price and delivery have been established, purchasers still want products that make them feel good about the work they do.

"Moral education," Alfred North Whitehead tells us, "is impossible apart from the habitual vision of greatness." The vision of greatness has been hard to come by within the disciplinary limits of industrial design. Any occupation needs heroes, but a profession ought to be able to provide its own. Industrial design thus far has not done this. Bucky Fuller cuts across all lines, but the designer's heroes are for the most part men whose background and discipline lie somewhere else, even if some of their achievements lie in industrial design. The great architects are the heroes of industrial design as they are of architecture. It is as though the Anglican church had to look to Rome for all its saints.

Few industrial designers other than Bel Geddes and Raymond Loewy have ever been famous, and Bel Geddes and Loewy were personally colorful romantic extroverts. Raymond Loewy was the most famous industrial designer, and this was by design. During the 1939 New York World's Fair, Loewy met Betty Reese, an ex-dancer who was doing publicity for the Fair. He invited her to become his press agent once the Fair was over. She did, and throughout a very successful career continued to describe herself as a "press agent" while her colleagues were all promoting themselves to "public relations counselors." From the beginning Ms. Reese operated with a direct intelligence. "What is your objective?" she asked Loewy.

"To be on the cover of *Time,*" he said.

"It will take ten years," she said.

It took almost that long. During that time and the ten years following, Loewy came closer than any other industrial designer to becoming a household word—although a household word that no household except his own could spell. It was not all Ms. Reese's doing. Handsome, suave, and French, Loewy was a natural, with an instinct for what the American market wanted and was ready for. An engineer by training, he was especially attracted to transportation vehicles and designed both automobiles and locomotives. He also designed a great many packages, including the Lucky Strike cigarette package—one of the best known single packaging designs in the history of the business.

In 1940 Loewy was summoned to the offices of George Washington Hill, president of the American Tobacco Company. What was needed, Hill explained, was a new design for Lucky Strike. The fee, Loewy explained, would be $50,000.

"If I pay that much, when will I get the design?" Hill asked.

"Oh, one nice spring day I'll feel like sitting down and doing it and you'll get your design then."

Whether it happened that way or not—Loewy steadfastly insisted that it had—he did design a new graphic treatment for Lucky Strike. It featured a bull's-eye on each side of the package, recognizable, Loewy boasted, no matter how it was set on a table or dropped into a gutter. Sales reportedly zoomed. Hill, however, was personally disappointed. He had always wanted a white package and later got one (also designed by Loewy) with the salable excuse that "Lucky Strike Green Has Gone to War."

Despite his engineering background, Loewy himself remained a stylist. And since styling was what both industry and the public wanted from design, this did not seem to limit him. It was not until William Snaith became managing partner that the Loewy office expanded its activities to include a range of services that had heretofore not been classed as design at all. These included retail store planning and marketing studies that were not directly concerned with designed products.

"If somebody comes to us for a bread package," Snaith would

An industrial designer and his designs make the cover of *Time*.

say, "maybe the best thing I can do for him is persuade him to get out of the bread business." In the fifties that sounded both heretical and nonsensical. But it was one of the directions that industrial design was taking. Richard Latham, one of the many Loewy alumni who left to start offices of their own, soon found that as an independent consultant he was spending most of his time *talking* to clients. And what he was talking about was not products necessarily, but planning. He was helping clients decide what they ought to do. An impressive number of them did it, and found that the advice was good. One client, the head of the radio and television division of General Electric, credited Latham with "moving our entire line ten percent ahead of where it would be otherwise." Latham long ago stopped doing product design and now spends all of his time talking to clients as a management consultant and planner.

The designers mentioned above were all "consultants," which in design merely means independent. Consultant designers are, by and large, not so much consulted as given assignments to perform. They are consultants only in the sense that the corporations they work for are clients who pay them retainers and fees rather than employers who pay them salaries.

When industrial design was new, this was a logical mode of operation. Except for some "art industries"—such as furniture, textiles, ceramics, printing—there was no reason for a firm to have a designer on the payroll. There just was not much product designing to do, and no awareness of design possibilities other than products. But design became a competitive necessity, corporations that once had no industrial designers began setting up in-house design staffs. The arrangement appeared to be less expensive than using consultant designers and more efficient, for staff designers understood at first hand the particular problems of the companies they served.

One drawback was that they knew the problems *too* well and were thus lacking in the fresh perspective that a consultant could bring to each project. Instead of putting consultant designers out of business, the presence of a corporate design staff sometimes actually enhanced the consultant's operation. Some design departments were set up by the firm's consultant designers, who were then retained for long-range planning services and for the less routinized projects. At one time Eliot Noyes served simultaneously as consultant director of design for IBM, Westinghouse, Mobil and Cummins Engine. Noyes's office designed some products for these corporations, but the bulk of his responsibility lay in identifying good designers and directing their work.

It was not uncommon for consultants to feel threatened by, resentful of, and superior to corporate staff designers, labeling them "captive designers." The implication was that anyone employed, rather than retained, by a corporation was not free to exercise his own imagination. The bitterness suggested by the "captive" tag was real. The American Society of Industrial Designers required for membership eligibility that a designer submit three different kinds of products. A member of the design staff of, say, General Motors or Ford was unlikely to have such credentials, and specialized designers were for many years not admitted to membership. As the trend toward in-house design staffs continued, the balance shifted.

The existence of professional societies was another indication of an emerging profession. If it had no Hippocrates and no Blackstone, design had at least created counterparts to the American Bar Association and the AMA. In the thirties the Industrial Designers Institute and the Society of Industrial Designers (later the American Society of Industrial Designers) were formed. In the sixties, after some 25 years of competitive rancor, the two societies merged to become the Industrial Designers Society of America.

The SID was founded for practical business reasons: New York State had introduced a tax law penalizing non-incorporated businesses. Professionals—doctors, lawyers, architects—did not have to pay the tax. Several designers banded together to fight, using Teague as a test case. The designers had their day in court, won, and Loewy appeared in Dreyfuss's office to complain indignantly, "Do you know what we've done? We've just made Walter Teague the world's first professional industrial designer."

Designers have some of the characteristics of both profession and business, but then so do doctors and lawyers. Although they sell their services, they are committed to maintaining their own professional standards when these come into conflict with others. A basic question in design ethics is whether a designer can honorably lend his talents and judgment to a product—such as cigarettes, high-sugar cereals or artificial sun-tanning equipment—that is harmful. The most common, and silliest, defense of such design is by analogy: a doctor, after all, is obliged to set the leg of a criminal because the oath of Hippocrates requires it. (If he does, however, his name is Mudd, after Doctor Samuel Mudd, who set the leg of John Wilkes Booth.) But the doctor in such a case is healing the person; he is not making criminality more effective, whereas if a designer improves and helps promote and sell a product he believes to be unsafe or just plain ugly, it is the evil itself that is enhanced by design.

These are not easy matters. A designer approached by a lamp manufacturer who wanted modest changes in what were, to the designer's eye, absolutely hideous lamps, said, "If I were to make them very much better, I might put the guy out of business. What do I do?" No client ought to be gifted with failure by a designer he has hired to bring him success. But if a designer really believes a product is atrocious, he can decline to participate in the atrocity.

The desperate search for professional status has led down some strange corridors. One of these is licensing. The license to design could, in the minds of some, cause industrial design to be respected as law, medicine and architecture are. But these fields are not licensed because they are professions; they are licensed because their reprehensible or inept practice would (and does) create a public hazard. Exterminators, security guards, notary publics, chiropractors, and street vendors are licensed for the same reason.

In the late fifties the ASID and IDI held a joint meeting at which, for the several hundredth time, the question of licensing was debated. One of the speakers expressed the hope that the question would at long last be laid to rest. It has not been. It is still debated today.

Designers debate a lot about education too, but I think they mean training. The fact is that neither license nor education is required for the practice of industrial design. Barbers and cab drivers must be licensed, but need not be educated. College professors need education but not licensing certification. Designers—like writers, artists, management consultants and corporate presidents—can practice without either. An ad for a book describing "200 rewarding careers that do *not* require a college degree" lists some of the top jobs for which no college preparation is necessary.

The four top ones are airline pilot, fashion designer, industrial designer, interior designer. Of those four, only the airline pilot is licensed.

Airline pilots require training that leads directly to the operations they perform in flying, and is compatible with the business of airlines. Designers are more likely to get training that has no immediate bearing on jobs open to them, and to develop values that are at odds with those of many employers. The disparity between campus and outside world, between academia and commerce, between ambition and achievement, is not peculiar to the universe of design, but it seems more dramatic there.

The proper student impulse is to redesign the universe. But since students don't have the universe as a client, it is difficult for them to get a purchase on the problems. Also, they don't know enough. But they do know that professional designers have failed to concern themselves enough with energy, economy, and values. Although crutches, wheelchairs, access ramps and special bathroom fixtures have been professionally designed, the problems of designing for the handicapped are more likely to be tackled in student projects than in design offices. Industrial design students are told by teachers and other saboteurs that the consumer needs an advocate, that technology needs an artist in industry, and that the industrial designer is both. They then land jobs (if they are lucky) in which they are expected to subordinate both consumer advocacy and art to the interests of marketing. Graphic design students are taught that theirs is a high calling, having to do with fidelity to form and balance, to color and line and proportion, only to discover that their clients and employers have no interest in any of that.

No one welcomes this phenomenon. It is perceived as an evil by everyone involved. Recommendations for correcting it are extreme and contradictory. On the one hand, design businessmen, or business design men, condemn the schools for not rubbing their students' heads in the "realities" of the world of commerce. This is reactionary and destructive, for it assumes that the realities of the business world are both grim and inevitable. What these hard-headed critics recommend is an enormous waste of the design potential—a waste that they are already implementing.

On the other hand, some of the most idealistic and charismatic design leaders—many of them teachers—tend to see the problem exclusively in visionary terms. (One decade's vision becomes another decade's routine. Teacher and designer Ken Isaacs built an exotic "living structure" in the fifties and people admired it from a distance, or laughed, or both. Since then it has been adopted and called a "loft," and can be bought in any large American city, or custom-made by cabinet makers anywhere.)

By the time the two industrial design societies merged in the sixties they had both become identified as "establishment." It was an uncomfortable time to win that particular recognition. Young designers and design students—like medical students and law students—protested against the design of instruments for war and defense, the design of products people didn't need, the design of unsafe products, the design of luxuries for the West when Third World nations were starving.

The designer ("the conscience of industry"?) appeared to be at best collusive, at worst gratuitously aggressive. A prominent industrial designer with a brewery account once explained, when asked about his relationship to consumers, "I'm the one who's gotta keep pumpin' that beer down their throats."

The student protests worked no miracles, but they have had a beneficial effect. While still fiercely protective of their own interests, the professional societies are less parochial than they might otherwise have been, and this is true of industrial designers generally, whether they are in or out of the professional societies. If industrial design is not an angelic profession, it is for the most part benign, and its potential for human service is vast and not yet deeply tapped.

The Way Things Mean

Love requires an Object,
But this varies so much,
Almost, I imagine,
Anything will do:
When I was a child, I
Loved a pumping engine,
Thought it every bit as
Beautiful as you. W. H. Auden

Sally said, "Look, Dick.
Jane can work, too.
She can make a boat.
A big, big boat."

"No, no," said Jane.
"This is not a boat.
This is a big red house."

Fun with Dick and Jane

Acting on the premise that anything can be made desirable if it can be made difficult to acquire, Tom Sawyer conned his friends into paying for the privilege of painting Aunt Polly's fence. The appeal of this scene is presumed to lie in the triumph of the con: Tom got the chore done, avoided working himself, pleased Aunt Polly, rewarded his friends with the satisfaction of measuring up, and got paid in the bargain.

For me the appeal has always resided in the goods he was paid in, the evocative list of precious objects: ". . . twelve marbles, part of a jew's-harp, a piece of blue bottle glass to look through, a spool cannon, a key that wouldn't unlock anything, a fragment of chalk, the glass stopper of a decanter, a tin soldier, a couple of tadpoles, six firecrackers, a kitten with only one eye, a brass doorknob, a dog collar but no dog, the handle of a knife, four pieces of orange peel and a dilapidated window sash."

Treasures all, but *why* are they treasures? Nostalgia? Maybe, but not for a day when such riches could be had. Every item on Tom's list is available today. What the list evokes is a time when things mattered. *Things!* In themselves.

In the preface to his selected poems Pablo Neruda writes of the desirability of looking closely at such objects as wheels, coal sacks, barrels and baskets because "from them flow the contacts of man with the earth . . . . In them one sees the confused impurity of the human condition . . . ."

"Those who shun the 'bad taste' of things," he warns, "will fall on their face in the snow."

Because we lay waste our energies with getting and spending and even with making and selling, just as Wordsworth said, we accuse ourselves of materialism. We should plead not guilty by reason of inanity: our obsession is not with things but with their appearances, their shadows, their connotations.

The world may be too much with us, but not the world of things. I think Americans today have little regard for the materials we live with (although we may appreciate their convenience) and an indifference bordering on contempt for the finished objects. Far from loving things too much, we are unable to love them at all, do

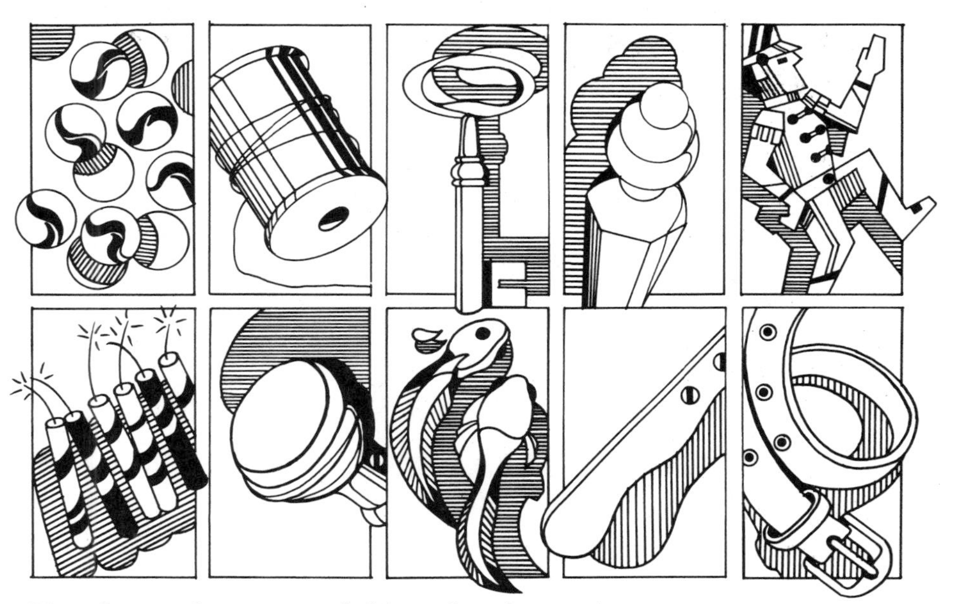

Tom Sawyer knew a good thing when he saw it.

not recall that objects once were cherished, do not remember how to cherish them ourselves.

And remembering is necessary. We are, as commentators will be telling us with increasing frequency, coming close to the year that has been famous for more than three decades as the title of a novel. At the beginning of *1984* Orwell's protagonist is about to commit a forbidden act: writing in a diary. The novel's ambience is set by Orwell's description of the diary itself.

> It was a peculiarly beautiful book. Its smooth creamy paper, a little yellowed by age, was of a kind that had not been manufactured for at least forty years past . . .
>
> . . . The pen was an archaic instrument, seldom used even for signatures, and he had procured one . . . simply because of a feeling that the beautiful creamy paper deserved to be written on with a real nib instead of being scratched with an ink pencil.

1984 takes place in a world where the sensuous feel of nib on rich paper means nothing; that is part of its terror. A book published in 1982, *The Meaning of Things,* by Mihaly Csikszentmihalyi and Eugene Rochberg-Halton, confirms the relationship between caring about things and caring about each other.

Paradoxically our inability to cherish objects is nourished by the technology and economy that provide objects in abundance. One could relate to a wooden table, for one had seen a tree or at least a lumber yard. But how does one relate to a Formica table top? Has *anyone* seen Formica? Does anyone know what it is? Is it anything? Does anyone care? The joy we sought (and sometimes even got) in the *ding an sich* has been supplanted by the satisfaction we build into (and sometimes even get out of) the symbol.

The symbol need not be real at any level (although it helps if it is tangible). When the attaché case first became a badge of success, or at least of clean employment, a New York pharmacy chain came out with a $4.99 knock-off described as "ALL IMPORTANT." Frankly "copied from an expensive $39.95 model," the case was advertised as "the symbol of the dashing, successful businessman-about-town." No mention was made of how much it would hold—in fact the illustration shows the case empty, in recognition of the fact that its function was not to carry documents, or even lunch, but impressions. The copy reads, "Aware of its growing importance . . . to the rising young executive, Whelan's copied an expensive all-leather model to bring you this handsome, budget-priced 'tachy' case with all the appeal of the higher priced model including the 'psychological lift' that comes with carrying one. . . . You'll be impressed with the impression it makes."

Even a fake attaché case is a real thing. But what a designer designs today may not be even a symbolic product but a system or part of a system, or a marketing program, or an exhibition on "ingenuity," or a feasibility study, or an identity program in which the subject identified is so nebulous that all form is arbitrary. And if it *is* a product, its character may be imparted as much by advertising as by the design.

We are, then, the un-proud non-possessors of objects whose chief substance is that of the transient symbol. Our Puritan fear of the love of things turns out to have been groundless after all, for we do not love things or even possess them: they pass through our lives as barium passes through the digestive tract, unassimilated, their function merely to flash signals along the way.

People used to trade what they had made for what somebody else had grown, or vice versa. Barter was the exchange of real stuff. Early money was intrinsically valuable before it became symbolically functional. Even the symbolic value of our money is twice removed, having been largely replaced by plastic. It isn't even necessary to carry the plastic either, for we can call in (toll free) the number that stands for the plastic that stands for the money.

The escape from the concrete is inseparable from a general indifference to essence. Hard hit by margarine advances, the dairy industry was advised by marketing consultants not to sell butter on the basis of its nourishment (which can be supplied synthetically) or its flavor (which many have lost the ability to discern), but to sell it instead on the basis of the status it confers upon the user.

In other words, the proof of the pudding is in the buying, an idea that has been carried so far it can hardly be parodied. No satirist is so wildly imaginative that his art is not endangered by the prospect of seeing his words become flesh, or at least polypropylene, at the hands of some manufacturer. The most caustic indictments have survived to become proud claims. A long time ago Lewis Mumford wrote that "in a society that knows no other ideals, spending becomes the chief source of delight; finally it amounts to a social duty." What an outrageous exaggeration that must have seemed to those at whom it was aimed. Yet today you can hardly pick up a business magazine without finding similar statements never intended to be pejorative. As the photographic historian Judith Mara Gutman says, in a book called *Buying,* "the whole process of buying . . . determines our daily pace, dictates our nightly rhythm. . . .Buying structures our lives."

In a world that has forgotten how to taste, the most persuasive reason for anyone's eating butter is to demonstrate that he can afford it. This has become the most persuasive reason for acquiring almost anything, although ironically it is no longer necessary to afford it in the old-fashioned sense of having enough money to buy it. The commonplace observation that most car buyers are in fact car renters suggests an economy's tacit but vigorous acknowledgment that what it makes is not—in the traditional sense—worth having.

Artificial obsolescence, according to Sibyl Moholy-Nagy, "implies an inability to love the things that make an identification of man and world possible." This inability, I believe, is implied by many phenomena other than artificial obsolescence; but certainly obsolescence is part of the pattern.

A product becomes obsolete when another product has been developed that is so far superior as to make it disadvantageous to keep on using the old one. With the introduction of the transistor, all radios using tubes were rendered obsolete. Now, to anyone with a sense of the value of things this did not mean that you immediately smashed all tube radios; it meant that the next radio you bought would be solid state. (My father, who bought one of the first radios sold in our home town, refused for years to replace it on the grounds that, any year now, television would be invented. Like a stopped clock that will tell the correct time if you wait long enough, my father was eventually right.)

But technical obsolescence is reinforced, and maybe even supplanted, by styled obsolescence, which makes the products you own look old by regularly issuing products that look newer. The newer products may not perform any better—at least not in ways that you need—but they serve a public relations function: to buy the new is to be perceived as new yourself.

More important, to buy the new is to be demonstrably able to afford the new. This has always been an important consideration in fashion design. In *On Human Finery,* Professor Quentin Bell argues that clothes are chosen for rhetorical rather than esthetic reasons, to show the wearer's financial standing. While any particular cut may indicate wealth, the cut-of-the-minute is almost certain to; for, whatever the style, it betokens an ability to spend money right now, and advertises that you don't need to hold on to last year's clothing.

But, in an age of fast, competent copying, can the statements

Pat Oleszko is a gifted artist who uses costume and movement to design unprecedented situations where the onlookers become incorporated into the design. There is, unfortunately, only one Ms. Oleszko, but, as shown here, she sometimes dresses for three.

made by clothing be trusted to tell the truth? In *The Language of Clothes* Alison Lurie speaks of the "world crisis in Conspicuous Consumption" brought on by the threat that "it might actually become impossible for most of us to distinguish the very rich from the moderately rich or the merely well-off by looking at what they were wearing." The threat was averted when manufacturers realized that "a high-status garment need not be recognizably of better quality . . . it need only be recognizably more expensive," an objective accomplished by putting the label on the *outside.*

Industrial products are also susceptible to fashion. In the *Harvard Business Review* Dwight E. Robinson told management:

> The behavioral complex underlying all stylistic innovation—by this I mean all changes in design which are not purely the results of engineering advances—can conve-

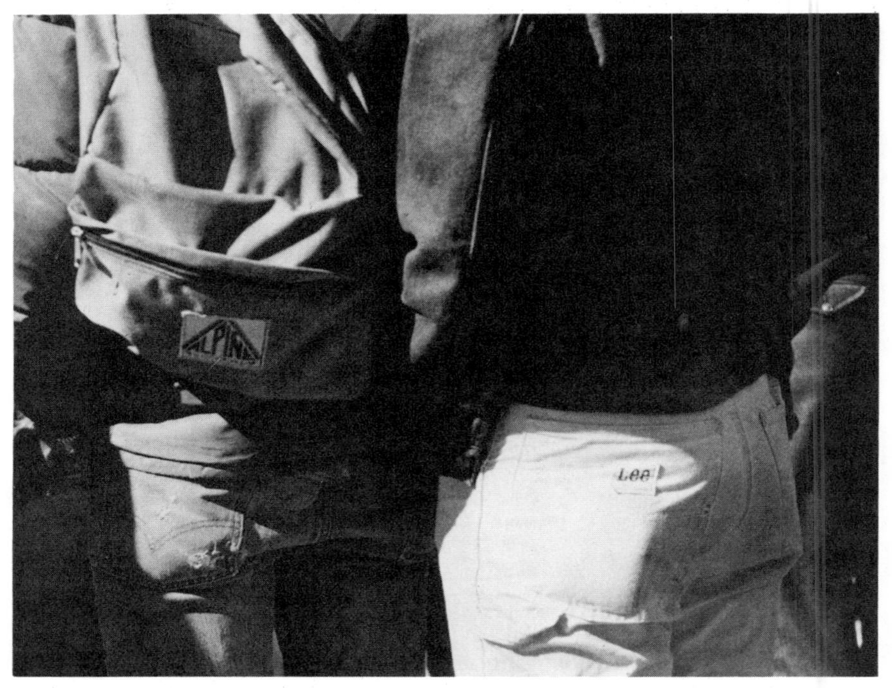

Labels on the outside of clothing were once in fashion only on the front of boxing trunks and the back of blue jeans. Today they widely advertise both the product and the wearer's view of himself.

> niently be summed up under the single word *fashion.* And fashion . . . is the pursuit of novelty for its own sake. Every market into which the consumer's fashion sense has insinuated itself is, by that very token, subject to this common, compelling need for unceasing change in the styling of its goods.
>
> The reason for this is that the stimuli of fashion derive solely from the *comparisons* that consumers draw between new designs and the old designs they replace.

Robinson's thesis is that the introduction of change by styling is essentially the same from industry to industry. Couturiers know this but other professionals either have not noticed or have been unwilling to acknowledge what they saw. Novelty in design, which is behind almost all styling, is not pursued quite for its own sake but for the sake of what it shows or says about a person. "People use cars to make a statement about themselves," American Motors chairman Paul Tippet tells us. Declarations have always been intrinsic to professional design. "The powerful Greeks did not need architects to shelter them," James Ackerman says, "but to communicate messages."

What messages? Messages about their money and their power, which is to say their status, which has more to do with public display than with possession.

There is a group of houses in a Long Island community, with gardens laid out like display goods on a counter: look but don't touch. This is not the standard competitive lawn care of the suburbs, redolent of county fair pie contests, with recipes for plant nourishment exchanged along with household tips about how to fight crabgrass. These gardens have nothing to do with gardening ability (which is commercially supplied) but with the status of possession. At one time the prevailing fashion there called for the display of large rocks, bought and hauled out at great expense. A friend of mine, upon being told that a rock he was looking at had cost seven hundred dollars, set himself up as a rock appraiser and strolled through the development calling out prices. "This one is a $450 rock," he would say, adding after a period of squinting and lip puckering, "at least." Dr. Johnson kicked a rock to refute Bishop Berkeley and establish the material nature of reality, but neither Johnson nor Berkeley foresaw a day when you could stub your toe on status.

The owners of these houses neither make the gardens nor want to. It is extremely doubtful that they enjoy them, or want to. What they want to do—want to be *perceived* as doing—is own them. But they don't own them either, and that is the whole point, for they have to keep buying them over and over again, like the favors of a paid companion. They think they are buying maintenance, but that is a technicality. Once you remove the pleasure of making things grow and the pleasure of seeing the results, there is nothing to a garden *but* maintenance, so the owners really are, in any meaningful sense, renters.

Turnover of property (as distinct from soil) has made this distinction necessary for years. *Scientific American*'s publisher, Gerard Piel, remarking on the turnover rate in housing, commented:

> Plainly, the so-called home owner is buying not a home but a housing service, much as he buys transportation, not a car, from the auto industry. His equity in these two utilities rarely controls before he turns in the old house or car for the new model. By the same token the total installment debt represents, from one year to the next, by far the major property interest in all of the other consumer durable goods in use in the country. The householder is correct in regarding these transactions as the purchase of a service rather than property. For the objects themselves are self-consuming, designed for depreciation to desuetude in 1,000 hours of service.

Piel may be correct in his identification of the householder's correct point of view, but it isn't a point of view householders necessarily hold. People like to think they at least one day *will* own the things they buy. "Once in my life I would like to own something outright before it's broken!" Willy Loman laments in *Death of a Salesman.* "They time those things. They time them so when you finally paid for them, they're used up." It would have been hard to convince him that he never had "it" paid for, that there was no "it," that he was not the owner of a refrigerator but a subscriber to a cooling service.

Installment buying is suitably geared to a society that does not care enough about things to want to own them outright. In any case, it is not the thing that matters, but what it represents, what—in one of the most tragically debased words in the language—is called its "image."

If industrial America ever had a loved object, there is no question about what the object was: the automobile. Americans used to fondle their cars, Simonize them (an all-day job), kick the tires, and sit meditatively on the baking running boards. The earliest automobile ads reveal that a car was sold on the basis of what it was like. Or what the manufacturer claimed it was like. Later, cars

The 1930 Marmon, designed by Walter Dorwin Teague.

were sold on the basis of what their manufacturers claimed they "said" about the owner, who was not expected to keep one for long because an old car, by definition, said all the wrong things. Now, under the pressures of foreign design and domestic despair, Detroit manufacturers, or their advertising agencies, are discovering "quality," as if it were an innovation.

But, in the meantime our most loved real man-made thing had been turned into an *un*real thing, before our very eyes. Not by *becoming* a symbol—as the Model T had—but by being designed as one.

Surely one reason for the present resistance to buying American cars is the increasing difficulty of feeling anything like affection for them. It would be simplistic and unfair to blame automobile styling for what has happened to Detroit. Yet it was considered reasonable and fair to credit styling with success when Detroit was experiencing success.

I think of my father's pleasure in certain cars. He is uninterested in cars as such, and is unmechanical to an extent that would make Jacques Tati look like an engineer. Convenience of repair was irrelevant, since no repair could be convenient enough for him to do. But he formed an attachment to his cars as they and he grew older. In the forties he drove a Model A Ford to work each day. When some high school kids craved it hungrily he sensed that their affection for it was akin to his own, and sold it to them for a few dollars. He began brooding immediately, and that night did not sleep at all. The following morning he tracked the kids down and bought it back.

Long before critics attacked the "American fat car" my father, who relates all experience to his wholesale grocery business, observed that Buicks reminded him of swollen cans of spoiled food. He was never tempted to buy a new car if the old one still served, but when he had to buy one he chose it on the basis of personal whim. He was the first on his block to own the 1946 Studebaker, and the first on any block to complain that the rear view mirror, which was set on the dashboard that year, afforded a superb view of nothing so much as the heads of the passengers in the back seat.

American automobile companies have always known of the need for a personal identification between car and driver, but they did not believe it was intrinsic to the car itself. They thought, and their research indicated, that the relationship could be lodged in the advertising and promotion. In February, 1982, a cougar attacked a nine-year-old boy in Pittsburgh. What was a cougar doing in Pittsburgh? Representing Lincoln-Mercury, because the designers of sales pitches for the Ford Motor Company still believe that driver satisfaction lies in the advertised association with savage power and sleek speed.

More than 20 years ago, when the American auto industry was wholly committed to big cars, when no oil shortage was threatened, when small foreign cars were publicly dismissed as a minor fad but worried about in the privacy of executive suites, when S.I. Hayakawa was not yet a senator or even a university president but merely our leading semanticist, *The Wall Street Journal* reported that "Ford Motors has called on the Institute for Motivational Research to find out why Americans buy foreign economy cars."

"The answer," Hayakawa responded in a magazine article, "is right there in the question." The Edsel flopped, he claimed, not because of errors in design or marketing technique but because the design and marketing were based on the findings of motivation researchers, who led auto makers to act on the working premise that the majority of the American population was mentally ill.

> Motivation researchers are those harlot social scientists who, in impressive psychoanalytic and/or sociological jargon, tell their clients what their clients want to hear, namely, that *appeals to human irrationality are likely to be far more profitable than appeals to rationality.* This doctrine appeals to moguls and would-be moguls of all times and places, because it implies that if you hold the key to people's irrationality, you can exploit and diddle them to your heart's content and be loved for it. . . .

The fallacy, according to Hayakawa, lies in neglecting to take into account that "*only* the psychotic and the gravely neurotic *act out* their irrationalities and their compensatory fantasies." The rest of us may, and probably do, entertain the fantasies, but we do not model our behavior on them. "Father may indeed see a bright-red convertible as a surrogate mistress and the hardtop as a combination wife-mistress," Hayakawa said, "but he settles for a lesser car than either because Chrissie is going to an orthodontist. . . .The more expensive an object is, the more its purchase compels the recognition of reality. The fact that irrationalities may drive people from Pall Mall cigarettes to Marlboro or vice versa proves little about what the average person is likely to do in selecting the most expensive object (other than a house) that he ever buys."

The motivation researchers Hayakawa describes so bitterly found new motivations wherever they looked. Thus a firm called Motivation Dynamics, Inc. announced at the beginning of the sixties that, "increasingly, American consumers are buying products for their own satisfaction and not to show off to their neighbors." Status was out, just as narrow ties were then. What replaced it? "Identity buying!" The consumer of the past bought "proofs and symbols. . . .In contrast, the identity buyer is seeking and responding to products and experiences which express, confirm and enrich his accepted new identity in his own eyes and the eyes of his family."

As illustration of this dramatic change, the researchers announced that "middle majority men and women want not radical departures from the past but products which help them reach smoothly the *next step in self-fulfillment,* be it a rose bush, a new sauce for the old meat loaf, a cake mix with a more refined brand image."

The trend in making, or faking, trends is endless. In case you were wondering whatever replaced "identity buying," *West Michigan Magazine*'s February 1982 issue reports an advertising agency's revelation of "integrity buying" as the behavior of the future. "Quality will become the new status," the agency's president says.

In other words, the more things stay the same, the more our rationalizations for them proclaim change. This is the message and mode of fashion design, which justifies arbitrary changes with claims of functional improvement.

When cuffs disappeared from men's trousers, fashion designers gave interviews explaining that the cuff was archaic and ill-suited to contemporary living. It collected dust, contributed nothing. When the trouser cuff returned, did it collect less dust and begin at last to make a contribution? Probably no fashion designer would argue the point; but the question never came up. Designers got rid of the cuff because there aren't many options for making trousers different. They restored it for the same reason. The pendulum swings, with a little help from its friends.

Neither designers nor manufacturers ought to be blamed for this (which is not to say they shouldn't be held responsible for their part in it). When they argue that they are giving people what they want, the argument is not entirely spurious. To be sure, what people want is conditioned by what they are schooled to want and by what they don't know about. Still, the public colludes in everything done to it, and probably always has.

When the American automobile industry argued (as it did for years) that to make cars safer was to cram expensive, unwanted features down the public's throat, like beer, they had a point of sorts. Their research and experience told them that speed and style were far more important to most car buyers than safety, and therefore speed and style, rather than safety, were what they designed for. (To hedge the bet, they argued that their cars were plenty safe.) Apart from the question of what their own moral position ought to have been, they seem to have been right about the public's position.

They apparently had been right about it for a very long time. In 1925 the inventor Charles Kettering presented at an American Chemical Society meeting a paper called "Motor Design and Fuel Economy." His paper argued that petroleum supplies might run short one day, a problem that could be addressed by the design of cars that would run on far less fuel. Kettering outlined the design features of such a car, which were essentially what they are today. He predicted, however, that it would be very hard to sell because it would not have "that reserve power so much desired by the motoring public."

To design objects on an unreal basis is to minimize our own reality. Travel has traditionally been "broadening," because it was an exercise in confrontation. You couldn't get from one place to another without experiencing the journey and the places. Now we travel vast distances sealed in chambers that are indistinguishable whether you are traveling from Cleveland to Detroit or from Boston to Yokohama. And to make sure no ambient reality seeps in, the airlines provide alcohol, music, and other drugs. Charles Lindbergh wondered, "Will men fly through the sky without seeing what I have seen, without feeling what I have felt?" We have gone further than he feared. People fly through the sky without feeling anything, without seeing anything, and without expecting design on the ground to offer anything different.

Down on the ground, two of the nation's most celebrated moral philosophers—the versifier Edgar A. Guest and the brothel madam Polly Adler—confirmed that a house is not a home. What these sages did not say (although I suspect Ms. Adler knew it) is that, for all practical purposes, a house is not even a house. It is an abstraction, conceptually manageable but seldom enjoyed in substance. Some architects speak of the house as a "spatial totality," but it is given to very few of us to live in one of those.

Instead we occupy what, for lack of a more pretentious term, might be called a "temporal partiality." That is, we know the house as a series of dynamic details: the flooring we pace, the knob we turn to open the door we slam, the plumbing we listen to. And probably that is how we experience any building.

Take New York City's Pan Am Building, for example. Its construction was justifiably feared as a horrendous architectural assault, a cheapening of the grandeur of Grand Central Station. Still, a couple of years after its vista-destroying erection, a group of us malcontents were grumbling about it when suddenly it occurred to me that only the day before, on the way to and from a lunch date, I had passed through the notorious edifice twice without even realizing it. Could a building so blandly encountered be the monstrosity we had been objecting to so strongly? Of course, what is monstrous is the callous intensification of the area's problems: too many people, too little sun. But these are routine city problems, and in its unapologetic contribution to their growth the Pan Am Building may be just a candid statement of urban design reality. In any case, we can cope with the building better than we had feared. It is less than the sum of its parts, for the parts are all we ever come in contact with, and standardization assures that most of them will not impinge upon our consciousness. (This suggests the way in which C. S. Lewis approached the problem of pain in Christian theology. Lewis reasoned that there is really no such thing as "all the suffering in the world," because no one person has to suffer it. The implications for architecture are consoling: perhaps there is no Pan Am Building, since no one person has to suffer all of it.)

Not only do we experience the whole of a building in terms of the part we happen to confront; we also tend to identify the part as the whole. To disguise the obviousness of what I am saying, I have coined the phrase *palpable synecdoche.* Here is an illustration of the phenomenon it describes: to a postman, the Seagram Building is really only the plaza and the bronze mailbox doors.

Detailing, then, is crucial. Yet this is where the architect is thwarted by product design. For at precisely this point—the vital detail—a building's designer is required to relinquish control, and specify rather than create.

These activities, to be sure, need not be mutually exclusive; indeed, our culture finds its origins in creative specification. The ultimate set of specifications is the first chapter of "Genesis," but that was in another country, and there was no concession made to the contractor. One cannot imagine the command: "Let the earth bring forth grass or the equivalent"—the standard wording of architectural specification. No, He specified with every assurance of getting exactly what was wanted. This divine avenue is not open to the contemporary architect, who may not wish to use a particular product design, but has to use it anyway, because nothing better is available on the market.

The market is the heart of the matter, for it is where we have to go in order to make mass production compatible not only with design excellence but with minority tastes. Our production technology has been a mixed blessing; but what other kind of blessing is there? The question is: how can the mixture be enriched? Perhaps the market can be made to yield more satisfactory answers than it has thus far, if only designers can put it out of their minds for awhile.

This, in an age when designers boast of being "market oriented," will not be easy. For the very nature of the product designer's role in industry tends to militate against his effectiveness. *He is schooled—and presumably motivated—to design things for people; but he is retained to design things for the market.*

This may seem like quibbling over words. It isn't. Words in this case are extremely important because they make it possible for us to conceal meaning. This is not necessarily conscious; but a designer may perpetrate a folly in the name of "the market" that he could not defend if he broke the market down into such concrete components as, say, his mother or his accountant. It is a semantic commonplace that, by contriving phrases that conjure up no precise images, we can avoid seeing the implications of what we are doing. Few manufacturers are so insensitive as to say (or even think): "I make products that are inefficient, unsafe, and ugly. People have bought them in the past without complaining much, why risk a change?" Instead they say: "In a highly competitive situation, design must operate at the level of consumer acceptance." Which—just look around you—frequently means the same thing. ("Acceptance!"—in other words, *literally* whatever the market will bear.)

Designers sometimes say defensively, "I don't design for the museums, I design for the marketplace." Why on earth should anyone design for either institution, when both are merely agencies for bringing what has been designed within reach of those it ought to have been designed for? No product designer *does* design for museums primarily, but a great many products are designed for the market, and for nothing and no one else.

When the designer concentrates his energies on the process of distribution and permits that to shape his work, his designs become less relevant to people. At the same time he is likely to become less immediately concerned with things. The product Willy Loman sells is never identified; it is simply the universal stuff of the marketplace. This omission takes on a significant negative weight in the play: what matters is the very fact that the product does not matter. The merchandise is subordinate to the process of buying and selling. The "market-oriented" designer is, like Willy, "a man way out there in the blue, riding on a smile and a shoeshine."

To design for the market creates a distance, sometimes unbridgeable, between the designer and the people who presumably inhabit the market. From that distance a designer's understanding of what people want comes chiefly from data supplied by the manufacturer.

Well, who knows better and on what better authority? There is a real dilemma for a professional who, after all, is retained not to express his personal integrity but to enhance his client's impersonal products. Since he is designing for people, isn't he obligated to design for what people want in the only sense in which anyone knows for sure what people want—what they will buy?

As with many hard questions, the answer is yes and no. People want lots of things. We want security and we want adventure; we want sensible, safe products and we want products that tickle our egos. To the extent that seemingly conflicting desires are compatible in form, both may be satisfied by design; in fact, to discover and express this compatibility is one of the designer's roles, and it is not a new one. (Even the surrey had non-functional fringe on top.)

Forklift truck designed by Niels Diffrient when he was a partner in the Henry Dreyfuss office. Design is often better in industrial work tools than in consumer goods, for the stakes are higher and the priorities clearer.

The problem is to distinguish between what is basic to a product's conception and what is incidental to it. An automobile is a device for getting people from place to place, just as a briefcase is a device for carrying papers. Both instruments offer a number of additional satisfactions. But if these extra functions are understood to be the main ones, it would be more efficient to make a product honestly designed to perform them, and do away with the extravagance of engines and file pockets. When a product's incidental value seems to be distorted out of all proportion to its basic value, the designer ought to reject the problem as both unworthy of him and beyond his talents. Design cannot help. If a person is indeed so insecure that a product has to be misleadingly designed in order to make him feel virile and socially acceptable, then what he needs is not motivational styling but psychiatry. He may as well take his Freud straight.

Design is not a democratic process, but a creative one. Democracy makes the designer responsible for appealing to people at their best with his best. It does not relieve him of the right to have ideas about what is worthy, nor does it absolve him from his duty to implement them by design. Are there products so undesirable as concepts that they defy "good design"? As William James taught us, the way to deal with a contradiction is to make a distinction. I cannot imagine a good torture chamber, although I know of some that have been very effectively designed. Weapons, as a rule, *are* well designed, which may tell us more about social values than about the design process.

The notion of designing a product that does nothing but confer status is preposterous when applied to automobiles and houses, but becomes less conspicuously wasteful and more conspicuously consumptive when applied to gifts. Each Christmas brings to the surface legions of objects that, like wampum and most other money, have no intrinsic value, but are media of exchange. While these are more noticeable at Christmas time, mail-order (or toll-free call) catalogs are yearlong resources for the acquisition of useless objects. The range in quality and price of useless objects is about as wide as that of useful objects. It includes at one extreme the mink toilet seat cover and at the other extreme the pet rock.

Nonsense products have spinoffs, just as real products do, for the only interest we have in either is novelty, which by definition cannot be sustained for long. A designer I know, in a moment of wild speculation, proposed a surefire get-rich-quick scheme: a casket for pet rocks. Someone unimaginative enough to buy a pet rock, he reasoned, would be unimaginative enough not to know how to get rid of it once he tired of it. A casket would provide the same short-term amusement the original product had, with the additional advantage of being self-liquidating.

From opulent boutiques to sleazy souvenir stands, there are in every American city stores that specialize in a commodity called "gifts." Towns too small to support barber shops or gas stations find room in the local economy for at least one gift shop, and the wares these shops sell are gifts in the purest sense.

A gift never used to be a gift until someone gave it, or at least bought it for giving. Up to that time it was a useful object with an independent value of its own. Today, however, we have a proliferation of gifts that are *designed* to be given. That is their function, and their form follows it boldly where angels fear to tread (with good reason: they may get kitchen timers stuck in their bellies). We all know that what counts is not the gift itself, but the spirit; and these gifts have the spirit designed right into them. A candid statement of this curious truth is the sign on a Great Barrington, Massachusetts, shop: "Designs for Giving." It is no great shakes as a pun, but it makes its point.

Caveat emptor presupposed a buyer who would handle the responsibility of protecting himself from shoddy products. To buy was to own and to own was to care about, and you couldn't trust someone else to care for you. There was no war between merchant and customer, but an uneasy truce based upon an honestly acknowledged difference of interest.

Today there is no uneasy truce but a cold war disguised by the advertised assertion that "we do it all for you." Customers would never believe that, but consumers may. That change is significant. A customer had personal stature. He paid his money and took his choice. He might even be a "tough customer." (Can anyone imagine a "tough consumer"?)

A consumer is not perceived as a person but as an economic entity, the last step in the distribution system, a device whose undignified, inhumane function is to use things up in order to make room for more things. A prominent package designer has stated that "the industrial designer . . . is of immediate value to his client only if he has the sensitivity and skill to move—not merely motivate—the consumer at every point of the way." *She* (still the pronoun of choice when advertisers speak of consumers) is a receptacle, an enormous trash basket. Marjorie Morningstar's uncle boasted that he was a "gobbage pail." He is the ultimate consumer. One of Russell Baker's funniest columns, "Incompleat Consumer," documents his plight as a middle-aged man confronting "my own trifling powers of consumption."

"The consumer is not a moron," advertising man David Ogilvy insisted, "She is your wife." Apart from the patronizing sexism of his assertion, there is the even more patronizing act of damning the buying public with the faint praise of being "not moronic."

Where the customer was tough, the consumer was seen as defenseless. Hence "she" requires consumer advocacy, consumer clinics, consumer guidelines, consumer indexes, consumer protection. A magazine for social science teachers advertises "a high school consumer textbook . . . to help your students make consumer decisions that are right for *them*—now, next year, and throughout their lives."

Conservative columnist George F. Will writes:

> Unfortunately, many citizens today think of themselves primarily as consumers, and think government's primary duty is to facilitate enjoyable consumption. So liberals, rallied by "consumer advocate" Ralph Nader, advocate a "consumer protection agency." Conservatives respond by championing "consumer sovereignty." Both sides seem to agree about one thing: citizens should be regarded, primarily, as consumers, and policy should serve consumption . . .

Yet it is hard to avoid concluding that consumers need all the protection they can get. There are a lot of messages aimed at us, hurled with billions of dollars' worth of energy, talent, and air time. While most such activity is pursued from a motive of intense self-interest, some grows out of the loftiest philosophical soil. An editorial in an advertising and marketing magazine complained: "Consumers are under-consuming. They still haven't learned to consume, nor have they been taught or persuaded that consumption is a responsibility in life." (Perhaps it is a patriotic duty to outconsume the Soviet Union.)

Curiously, when it does not have to carry the missionary burden of converting us into born-again consumers, advertising can

Chemex instructional brochure by Peter Schlumbohm.

enhance the experience of product ownership. One of the most enjoyable aspects of buying over-the-counter pharmaceuticals used to be the propaganda that came with them. When you opened a tin of aspirin or cold remedy box, or a box with a shaving cream tube inside, you got an illustrated bulletin from the manufacturer, congratulating you on your selection. Such products were customarily opened in the bathroom, where the literature was read. The reading gave pleasure and information, and boosted confidence, as did the hangtags on hard goods.

One would suppose that the worse our products get, the better our product literature becomes, but this is not the case. When the design integrity of the Chemex coffeemaker was jealously guarded by its inventor, Peter Schlumbohm, each product was packed with an eccentric and instructive manual of coffeemaking and good living, supplemented by promotional copy for other Schlumbohm inventions. Now that the birch handles have been replaced by plastic, the text has itself taken on an impersonal plasticity. Few products today are accompanied by any literature at all. The rationalization for this is that there is no use selling somebody who has already bought. But the omission suggests that there isn't much to brag about.

During World War II, in the interest of speed and efficiency, magazines sent overseas were stripped of their advertising (G.I.s weren't buying much at the time) and reduced in size to dimensions that could be easily mailed in bulk. These were called "pony editions." You got the magazines much faster that way, but they were disappointing, without the richness of expensive stuff. Some servicemen wrote to *The New Yorker* to demand the real thing, ads and all, however late.

With so much of a product's value dependent upon how it is advertised, rather than upon how it is made or works, advertising itself becomes a value. In the days of customers, there was a clear distinction between product and sales message; certainly there was a clear distinction between customer and sales message. But in the age of consumers, the consumer often *is* the advertisement. At least he is the medium, which is reputed to be the message. There seems to be no limit to the number of people willing to turn themselves into advertisements, or to the number and kind of advertisements they are willing to carry.

When goods were more likely to be valued for themselves, the smuggled ad was one way of acquiring them, particularly for children. The giveaway was a form of implicit bribery with a *quid pro quo* understood by both sides. The free Pegasus insignia on a bicy-

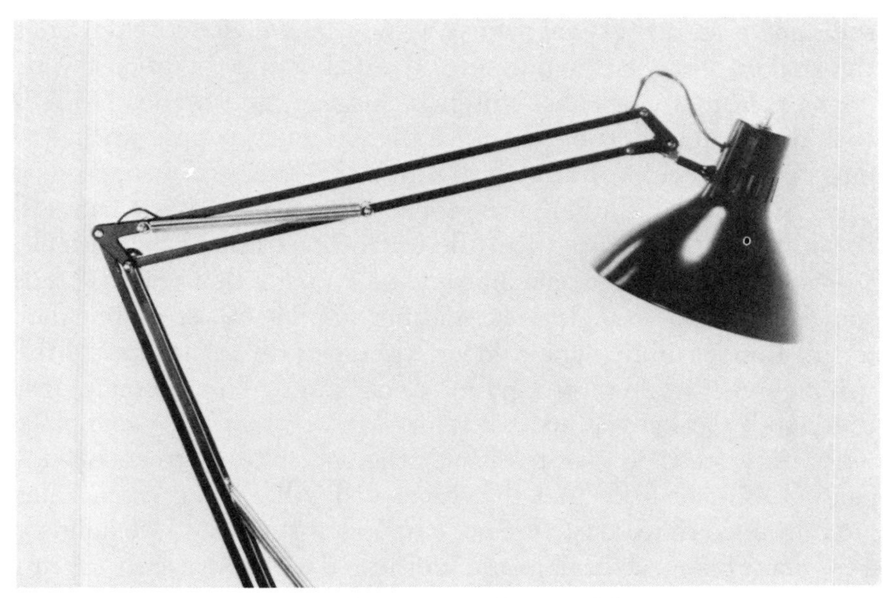

The Luxo lamp, a staple in designer's offices and homes.

cle gave the rider an attractive piece of junk to clutter his fender with and gave Sucony Mobil free advertising. A Tom Mix lariat proclaiming the owner one of the Ralston Cowboys elevated his morale while plugging the cereal.

There has long been a lively urge to advertise business enterprises by collecting book matches, or stealing restaurant ashtrays and hotel towels that advertised where you had been. It was an activity traditionally most attractive to high school students or to the kind of adults who buy whoopee cushions. In any case, the advertising message enhanced, rather than diminished, the product's worth. Sometimes there was a straight exchange of services: Mail Pouch ads appeared on barns, in return for which farmers got their barns painted (red) free of charge.

But why, after people have been sold a car, are they willing to ride around with the dealer's name and address prominently displayed on the back? Why don't customers refuse to have the dealer's sales pitch bolted to their bumper? Why do people carry luggage with designers' initials worked into the motif? (Only partly for the same reason they wear shirts with animal codes stitched onto the pocket, designed to announce the maker and the price.) Why do people order golf balls, golf bags, luggage, and highball glasses with airline insignias on them? There are even people willing to buy stainless steel airline flatware for home use, and I suppose, if they were available, there are people who would buy the back-of-the-seat trays too, along with the microwave-thawed Swiss steak.

Market research people are fond of insisting that your car tells something about you. It does, and not just in the sense that they mean. Besides telling who sold you the car it is likely to herald the vehicle's visit to Ausable Chasm, Cedar Point, Fort Ticonderoga, Storyland, Gaslight Village or its voyage on the Lake Champlain ferry. Any roadside merchant with the audacity to attach his unpaid ad to a car finds that the driver will willingly carry the sales pitch from coast to coast. We are a nation of shills, and we do it for free, permitting car dealers and clothing designers to turn us into ads. Some buyers are less passive, choosing particular tee shirts *because* of the sales message they carry.

Designers themselves retain their love of things (although surprisingly, they do not seem to be especially astute buyers) and ought to be encouraged to communicate and act upon it, to broadcast the pleasure in being "lovers of paint and metal." When asked to name outstandingly designed products, industrial designers are apt to take two lines. They cite either the "in" designer products—the Luxo lamp, appliances by Braun or Olivetti, furniture by Herman Miller or Knoll—or the "chic humble" class of generic objects: the safety pin, the rolling pin, the needle, the egg.

The lack of satisfaction in our manufactured goods has a positive side: it is a motivating force behind the currently fashionable do-it-yourself activities. But baking bread, weaving cloth, and growing vegetables, however desirable in themselves, are no solution to the problem of goods that do not satisfy. We still must learn to make the resources of technology yield objects we respect and love, objects designed for use and affection rather than for sales and acceptance.

Or we'll fall on our face in the snow.

The Prop

Goldilocks sat in the great big chair. It was too hard. She sat in the middle size chair. It was too soft. She sat in the baby chair. It was just right— but it broke when she sat on it. Goldilocks and the Three Bears

That capsule history of chair design is still an accurate description of the state of the art. Does it matter? It matters very much to designers, who adore chairs to the point where they sometimes worship at their feet, and to design students, for whom chair design is a standard advanced problem. But the rest of us apparently also take chairs seriously. The billions of dollars spent on them each year run into double digits. If the world is a stage, then all designers are set designers. The chair is the minimal daily prop.

It is a curious basic artifact, for it has nothing to do with basic survival needs: food, shelter, clothing. There has to my knowledge

Barcelona Chair designed by Ludwig Mies van der Rohe, 1929.

never been a world chair shortage, nor has the need for additional chairs been established by the Surgeon General. Yet legions of designers go on designing them. Although no designer is so maladroit and inept that he cannot make a satisfactory chair, few have created chairs that are handsome, sound, comfortable, and not injurious to health. When a chair is original in design concept, it can make a designer's reputation. If it is possible to sit in it with anything less than acute pain, this is a bonus.

Architects in particular find chairs challenging: Mies van der Rohe, Eero Saarinen, Marcel Breuer, Le Corbusier, Frank Lloyd Wright and Alvar Aalto are among the celebrated architects who searched for the right chair solution like biologists hunting for a cancer cure. They are not alone in endowing chairs with cosmic significance. A standard exercise in traditional philosophical inquiry has to do with determining whether or not a given chair really exists. Why a *chair*? Henry Thoreau lived at Walden Pond under the most stripped down, spartan circumstances in our history, in a cabin bare of any excess. To make the point, he equipped it with three chairs: one for solitude, two for friendship, three for society. (Comfort did not concern him any more than it concerns designers.)

The chair dominates our language as it dominates our environment. Our ways change, and with them our diction, but whether a group has a chairman or a chairperson, the symbol lingers on. Folksy: pull up a chair. Formal: please be seated, Mr. Abercrombie will see you soon. Slangy: park your carcass.

My father distrusted certain novels and movies because in them people invariably said, "Won't you sit down?" My father didn't believe people ever really said that, except in books or movies, and in his experience no one did say it. On the other hand he accepted "sit, sit" as a routine bid for environmental control disguised as hospitality.

The chair recognizes you, and we recognize the chair. During the Watergate hearings the Department of Justice wondered publicly whether it was legally possible to arrest a sitting president. One of the highest university distinctions is to hold an academic chair. The witness in a courtroom trial is seated high enough to be displayed to the jury and spectators, yet low enough to be subordinate to the judge. The placement of chairs is one of our oldest forms of gamesmanship. Has there ever been a job interview in which the candidate sat *above* the interviewer?

The role of the chair as an element in human interaction is relatively unstudied. Not to worry. What we need is not research but awareness. You can easily do informal studies of your own. Try taking Polaroid pictures of your living room before and after a party. Many rooms will look quite different, but some will look pretty much the way they always do, except for the addition of peanut shells, glasses, and filled ashtrays. It depends a lot on the kind of party giver you are. If the furniture has been radically dislocated, the normal configuration of your furniture is presumably not the same as that for a party. But what if the furniture is *not* out of place? Does this mean that you are *always* ready for a party? That the furniture is too large to move? That you gave a dull, or at least remarkably static, party? Or that it was a standup party?

In the last case, it is worth asking *why* it was a standup party. "No, thanks, I'd rather stand" may be a revealing statement. There are social occasions in which guests are afraid to use the chairs, just as presidential candidates Carter and Ford during their first television debate were afraid to use the bucket seats that forlornly stood beside them. *Even during the 27-minute audio failure, neither candidate sat down.* To be sure, sitting down would have been a cumbersome process because of the cords running from their throat mikes to the audio equipment. But sitting down is a position of trust, like turning one's back; and these were two men not in a position to trust each other or themselves. What if sitting were perceived as a sign of fatigue? Would the nation trust its government to a man who needed a rest? If Ford *had* sat he would have lost the advantage of being taller than Carter. Also, to sit down would mean having to get up, not always easy to do gracefully.

Such concerns, dramatized in a national event that was otherwise devoid of dramatic interest, affect us almost every day. We have all been in rooms in which some people dominate the conversation because of who they are, but also because of where and how they sit. We have all seen (and been) people unable to get into a conversation because of where they have managed to place themselves. A very deep, large lounge chair, for example, is a good place from which to deliver nuggets of wisdom if you are being lionized at a gathering, but a very bad place from which to try to get a word in edgewise if no one is especially eager to hear you. We have all seen, or been, people stuck wherever they lighted or were steered into by an obliging hostess. It is impossible to furnish a room intelligently without taking this kind of thing into account, and that is one reason for the variety of chairs we seem to need.

To some extent, the placement of chairs is not entirely our doing. Products contribute to the design of other products. The

White sliced bread, a universal symbol of cultural sterility, has nevertheless engendered a plethora of accessories.

process may be circular: think of the toaster, designed to accommodate the white sliced bread which is itself designed less for flavor or nutrition than to fit into the toaster, which was designed to . . .

Products contribute also to the design of the spaces in which they are used. In this respect television is a powerful design influence. Until its advent there were very few living rooms in which the chairs faced the same way. Chairs used to face other chairs, or a sofa, so that people could talk to or look at each other while they listened to the radio. Chairs were positioned near lamps for reading. But television has transformed the living room into a screening room, and the design of chairs for viewing is not the same as the design of chairs for reading or thinking. With the current fashion of using commercial objects in the home, we can expect any day now to see living rooms equipped with entire rows of tandem movie house seats. (In the spirit of "high tech" some designers already use these, as Hollywood personalities have for years.)

The prospect of physical discomfort has not deterred anyone from buying, or sitting in, chairs that hurt. A painful chair, however, is more willingly bought and endured if it carries the imprimatur of a museum or some other respectable design authenticator. Randall Jarrell noted, with great wit but no exaggeration, that there are people who ". . . will sit on a porcupine if you first exhibit it at the Museum of Modern Art and say that it is a chair. In fact, there is nothing, nothing in the whole world that someone won't buy and sit in if you tell him that it is a chair . . ."

Here are some things we can sit on: rocks, tables, beds, floors, the ground, curbs, packing crates, fences, tombstones, logs . . . and hundreds of thousands of existing chairs. Yet chairs proliferate as if we needed them, and for every one freshly displayed in a furniture showroom there are at least a hundred on the drawing boards of young designers poised for stardom.

Classic chairs (you may not have realized there are "classic" chairs, but to designers there are even "classic" panty hose displays and beer cans) are sometimes stubbornly functional. In one of the myriad Bauhaus declarations, Walter Gropius wrote: "In order to create something that functions properly—a container, a chair, a house—its essence has to be fully explored for it should serve its purpose to perfection, i.e. it should fulfill its function practically and should be durable, inexpensive and 'beautiful.' "

As with all counsel of perfection, the advice is not especially useful. That is, no one would argue that a chair should *not* fulfill its function practically, that it should not be durable, or that it should be costly or ugly. But somebody might argue that it is unnecessary and pretentious to explore the essence every time. In fact, somebody *has* argued that. Namely, C. Northcote Parkinson, the author of *Parkinson's Law,* a book so disarmingly funny that we live in mortal danger of ignoring how serious it is. "Life is too short," Parkinson said. "When asked to design a chair, the designer shouldn't sit down and gaze at the sky, saying 'What is a chair? What are the *elements* of the problem? . . . What is the true philosophy of chairmaking?' It all takes too long, and costs too much, and the result is horrible anyway. Better to agree together on what a chair is. At the end of it, one designer will obviously be better than another."

Parkinson wasn't suggesting that designers already know what a chair is all about. He was suggesting that designers don't know but ought to find out, ought to establish some consensus on what a chair means, so that design could proceed as a professional discipline—a professional discipline being one that has some body of knowledge upon which there is qualified agreement.

But the function of a chair is not all that easy to establish. Some years ago a friend of mine praised a chair she owned with the remark, "It is a very good chair for sitting." At the time that seemed a strange feature to single out. Sitting, I thought, is what all chairs are for. But it isn't. The function of a chair may be to fill a corner, dress up a room, keep a table from looking unattended, organize space, impress people, depress people, intimidate or welcome people.

Even when the function of a chair *is* sitting, there is considerable latitude in how that requirement is to be met: a chair may be designed for elegant sitting, for long-term sitting, for brief sitting, for comfortable sitting, for calculatedly uncomfortable sitting, for "seating," which means uncomfortable sitting in large groups. Even chairs designed primarily for sitting obviously are used for many other things. Exercise. Play. Drying clothes. Storing objects. Making love. Standing on to change light bulbs.

The true function of a chair is likely to be symbolic. If a man's home is his castle, a man's chair is his throne. (A woman's chair still has fewer *ex cathedra* implications, unless she is elderly.) The gospel according to Archie Bunker is most authoritatively delivered from the nerve center of the household: his easy chair. The canvas-backed director's chair is practical in that it can be folded up and moved from studio set to location, but that is not its real practicality. After all, the film industry spends millions of dollars moving heavy equipment around all the time and could haul chairs as easily as it hauls cameras and cranes. The function of the movie director's chair is to carry the name of the director.

Because only the powerful had them at first, chairs—unlike, say, cars, watches and Gucci shoes—may actually have *begun* life as status symbols. Advertising has used status chairs to consistent good advantage. "As long as you're up, get me a Grants" may have been the most casual demand for service since Mae West ordered Beulah to peel her a grape. The tone of the ad was good humored snobbery, with the speaker issuing his command from a chair that instantly instructed readers in the power and affluence associated with the whiskey.

Our plainest chair symbols are the so-called modern classics—the Barcelona chair, the Wassily and Cesca chairs, the Eames leather lounge chair—loved by designers and used in their offices and in those of their clients. But the most deeply symbolic chairs probably are those that Goldilocks discovered, and they probably tell us as much about human factors in chair design as most designers know.

The story of *Goldilocks and the Three Bears* expresses a close relationship between person and chair: "The moment they stepped into the house they saw that someone had been there. 'Umph!' said the papa bear in his great big voice, 'Someone has been sitting in my chair' "—suggesting a kind of rape, an invasion of something too highly personal to be shared. The violation was instantly apparent. Bears, generations before Esalen or est, understood the sanctity of space.

Even symbolic chairs have to be sat in, and sitting is itself symbolic. Sitting is not plopping. It is a muscular movement, and can be performed gracefully or brutally. Designers sometimes question whether they ought to design for the way people sit or the way people ought to sit; chair design, however, is usually concentrated on neither, but on how chairs *look*.

The literature of design is extremely revealing in this regard. There are plenty of books about chairs. In addition to them, museum chair collections are described in lavish catalogs. And if you add to these all of the commentary on chairs in books that are not about chairs as such, you get what amounts to a substantial library dealing with the design of a single object. Some of our best minds have taken on the task of assessing individual chairs. It is instructive to read these critiques both for what they say and for what they do not say. You will find the chair discussed as a problem in appropriate form, as you would expect. You will find the chair discussed

The first function of seating is to support the human fundament. Some seating is designed to support it only minimally. If counter seats were more comfortable, people might linger.

at great length as a problem in materials, as you would also expect. You will find the chair discussed as a problem of architecture—that is, the creation of an independent structure from a variety of materials. You will find the chair discussed as a problem in esthetic statement. You will find it discussed in terms of the effect chairs have on spaces.

But very rarely will you find the chair discussed as a problem in supporting the human fundament. That is no minor oversight. Nothing tells us more than this consistent omission, although precisely *what* it tells us is something we are likely to know already from sitting in chairs that designers have designed. Trade magazines in recent years have heralded a "revolution in ergonomics," a singularly ugly British term to describe what in this country is known by a slightly more euphonious, but figuratively even uglier, term: *human engineering.* Whether it's called ergonomics or human engineering or human factors, the activity itself is nothing more than an attempt to design a product in a way that takes into account the characteristics of the person who will use the product. That is hardly a revolutionary notion. The first man or woman to sit on a rock may not have been conscious of ergonomics, but the first one to turn the rock around until he found its most comfortable surface was. But in the time that has elapsed since then, designing a chair to satisfy the requirements of the human anatomy has not become the prime objective in designing, manufacturing and selling chairs, although it has become an increasingly popular theme in chair advertising.

When structure is discussed in books on chair design it is almost invariably the chair's structure, not the sitter's structure. Designers who will go to any investigative length to learn about materials neglect such elementary materials as flesh and blood —although the flesh is weak and the flow of blood has been greatly impeded by a succession of prize-winning chairs.

To design a chair without knowing how the body is designed would be like designing an egg carton without knowing how eggs are designed. But although probably no package designer would commit the latter error, chair designers customarily commit the former. And they get away with it, because human beings are more forgiving than eggs. An egg is powerless to compensate for bad design, and will simply break under the strain. But human beings can adjust to the chair's inadequacies. Since the chair will not accommodate us, we get bent out of shape to accommodate it, turning, twisting, scrunching up (sometimes with the aid of pillows or books) until we find a position that is tolerable.

Is there a solution? No. That is just the point. There can't be *a* solution, because there is no single body. Human bodies vary vastly with age, gender, diet, traumatic history and genetic inheritance. Moreover, living bodies can't be trusted to hold still; they tend to *move* from time to time. Clearly, no one chair will do. All the more reason for knowing as much as we can about how bodies work.

By observing how we use chairs we get some clues as to how they might be designed. We can sit in a chair or on it. Sitting in it will, all other things being equal, give us more support. But all other things are not equal, and more is not always better. We need better support than most chairs give us, usually in the lumbar region, for reasons having to do with ergonomic errors made when people were designed, and aggravated by us ever since.

Chairs support more than bodies; they support body language. Sitting *on* a chair, which in many cases means sitting on the edge, is both a literal and a figurative position. In the case of a horror film, the body's posture expresses the mind's fear. Scared stiff, the body is poised for attack or flight. The more highly participatory a situation is, or promises to be, the closer to the edge of the seat one moves. Benched athletes sit near the edge of the bench, (which, if backless, supports only their bottoms in any case) in the bodily expectation, however unrealistic, of being sent in, and in the excitement of watching the game. Their body language expresses a commitment to whatever is being played out in front of them.

But our bodies, like ourselves, may fear commitment, and with reason. To occupy the edge of a chair in a sedate living room does not express an urge to leap into the fray but a commitment withheld from the chair. Again, the lumbar region knows more than we do about what chairs are for. If the seat of the chair is too long for the seat of the sitter—a common failing—there is no lumbar support. To gain it, short people slide their buttocks back to the point where their feet come close to losing contact with the floor. Persons thus situated, unable either to keep their feet on the ground or their heads in the clouds, are disturbingly nowhere. To make matters worse, with the body thus cradled in the chair, the mind jumps ahead to the problem of how to eventually get out of the damned thing.

We need chairs that let us sit comfortably on their edges and deep in their interiors, and that do not assign our legs to a tenured position. We are getting some such chairs, and will get more, for the cure is partly rooted in the malady. This may be a cruel source of optimism, but I take great hope in the fact that a designer's body is as vulnerable to misuse as anyone else's. The longer and oftener designers sit in the chairs they design, the more likely they are to develop back trouble, which turns out to be the strongest possible incentive for learning about backs. The redoubtable Ward Bennett, a distinguished designer of many things, including more than a hundred chairs, acknowledges, "My interest in chairs stems from a back injury."

Motivation is just the beginning. As Bennett discovered, it isn't easy to get reliable information about backs and sitting. A designer who has helped make it easier is Niels Diffrient who, when he was a principal in the Henry Dreyfuss office, helped develop anthropomorphic information to support design. Diffrient's interest in body mechanics is reflected in the chairs and seating he has designed, which includes seating for American Airlines.

How an airplane seat is designed is extremely important, because the passenger is literally captive in it, his captivity reinforced by illuminated signs and robotic messages from flight attendants. When a plane lands, passengers rise from their seats. This activates the senior flight attendant, who explains that, since the plane has not yet reached the gate for "deplaning," the time is not yet ripe for deseating. Passengers already know this, having heard it all before. They stand anyway. They know that in most cases there is no use hurrying, for they are locked into a sequence of slow stages that cannot be speeded up. Nevertheless, fully aware of all this, rational people like you and me *still* spring up out of our seats, can hardly bear to stay in them. Surely this is psychological: the flight is over, we made it, we want to be on our way. But just as surely there is a bodily urge to escape from the prison of airline seating.

The seats are in fact designed to perform well, given their functional priorities. They are not load-bearing devices primarily; they are fare-bearing devices. To be sure, they have to satisfy minimal passenger comfort demands (just as they have to satisfy minimal F.A.A. safety requirements), but their real business is the temporary storage of ticketed bodies, and that they do with an ever increasingly efficient use of space.

The mission of an airline is to get people from one place to another. That in theory is the service they are selling. But since no airline accomplishes this any better or faster than any other, airline competition and charges are based on what they call "service." This includes food, drink, high-fructose music and other distractions. An industry so earnestly geared to distractions has a hard time taking design seriously, and there is in all commercial air travel a relentless concern with everything but flight. This may have the effect of putting all emphasis where it normally does not belong. I wonder, for example, whether this dedication to irrelevance explains why airline announcements always place the emphasis on the least important word. ("Please remain in your seats until the captain *has* turned off the seatbelt sign," "Please wait until we have taxied *to* the gate.")

American Airlines seats on some flights provide the extra support the lumbar region needs. And because not everyone needs the same amount of support, the seat backs are inflatable. The airline I use most often doesn't know or care about my lumbar region, or my ischial tuberosities either, for that matter. Furthermore, to achieve a comfortable lounge appearance, its seats have a unit of extra padding at the top, which pushes the head forward. Now it happens that your head sometimes wants to go forward, but only if it has the option of coming back, which the airline seat does not give it. So your head is not only pushed forward, it is held there until you reach Grand Rapids.

Staying in one position for too long, which certain chairs encourage, is an insult to the circulatory system, as alcohol is to the liver. A doctor at St. Vincent's Hospital in New York's Greenwich Village told me that, as a public service, when he comes across bums sleeping off drunks or highs, he routinely turns them over to prevent bedsores!

The problem of comfort, of designing a chair with a clear no-

President John F. Kennedy in his prescribed chair.

tion of anatomical and circulatory needs, is occasionally solved in chairs you might not be comfortable living with because of how they look. Others try to achieve comfort by *duplicating* the contours of the body, which is like using ginseng to restore potency, because of what the root looks like. There are even upholstered chairs designed to duplicate the *composition* of the body. With their frame and springs and stuffing and fabric or leather covering, these chairs are, like us, made up pretty much of skin, bones, wrinkles and a little fat.

As for that "ergonomic revolution" hailed in the trade press in the 1970s, it had been saluted before, in the 1860s, when a trade magazine advised that "comfort, convenience and adaptation to health are the chief ends to be secured in the construction of a seat." In 1876 the Wilson Adjustable Chair Manufacturing Company in New York advertised a chair designed to alleviate the pains that sedentary workers feel in the small of the back.

Why does this have to be relearned 100 years later? Santayana taught us that those who do not know history are condemned to

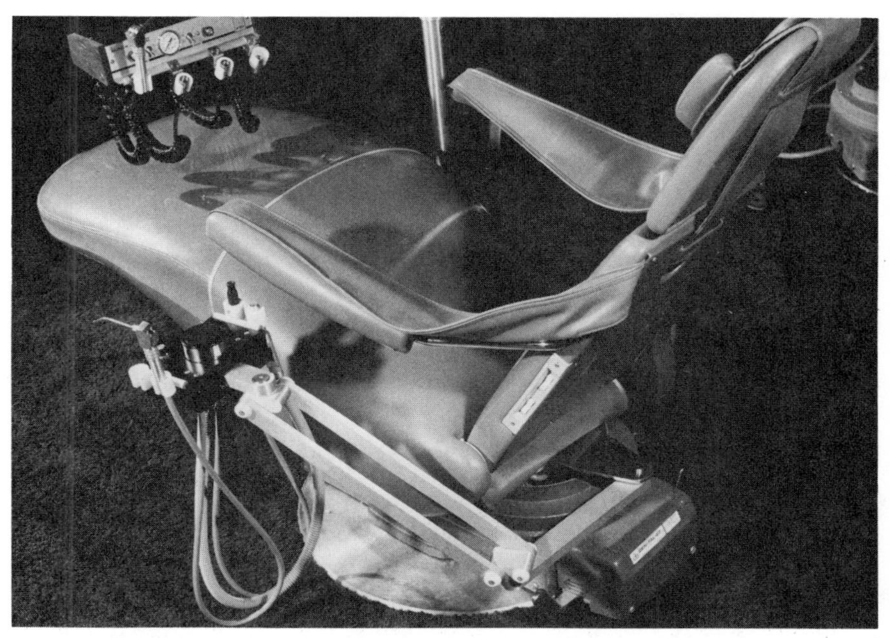

Dental-chair design makes the patient accessible to the dentist.

repeat it. That surely is true in design as in anything else, but in design there is a corollary: those who do know history are privileged to repeat it at a profit.

Some of this is fashion, but—as with all fashion—it may go deeper than anyone means it to. "Fashion," Kennedy Fraser tells us, "isn't simply hems and feathered hats . . . it is connected very, very closely with how we live." So the periodic reintroduction of rocking chairs may go beyond marketing whimsy. Although rockers are customarily associated with old ladies, in fact, children love them and John F. Kennedy, our most vigorous president since Theodore Roosevelt, was strongly associated with a rocker, in his case as a prescribed therapeutic device for a diseased spine.

One theory holds that the rocking chair gained popularity among the early Americans because it institutionalized American informality. Americans liked to lean back in their chairs, balancing on two legs, and put their feet up on tables. The rocking chair made such behavior acceptable, or at least safe.

The rocker's essential form makes bentwood the natural material for it. One might wonder whether the Shakers' excellence in designing and making bentwood chairs that move can be traced to their inclination to shake. Probably not. It seems more reasonably attributable to the sense of proportion they brought to all their designs, for the rocker is an unusual challenge in proportion. It is easy to mount a chair on rockers, but difficult to keep the result from looking preposterous.

While chairs for general use may have suffered from insufficient attention to function, this is far less true of chairs for special use. If you are working a farm, milking a cow, fitting someone with shoes, or getting a haircut or a root canal job, you need a particular kind of seat and there is a good chance of its functioning well. For someone. What the dentist's chair, the barber's chair and the electric chair have in common is not just that they are all seats we would like to avoid, but that the end user is not the person sitting in the chair but the person operating it. The human need to be met here is the need of the barber for an accessible head and neck, the need of the dentist to see and to work inside the mouth, and the need of both to keep the customer's arms from interfering with the task at hand.

More recently, refinements have been introduced. Now that haircuts cost $25 instead of 25¢, what the customer buys is not just the haircut itself, but the experience, including the experience of sitting in a chair that is not only comfortable but at times downright sumptuous. Similarly, dentists have shifted their attention from the mere pulling or filling of teeth to a wide range of services, including blood pressure checks and smile analysis. Psychology becomes important, at least the mention of psychology does. I went for years to a dentist who had co-written a book called *Psychodynamics in Dental Practice,* addressed in part to the relationship between the dentist's chair and the patient's anxiety. Unfortunately (for he was an excellent dentist) he retired early, from back trouble brought on by bending over chairs.

Special purpose chairs do not always retain their special purposes. The original version of the Hardoy, or butterfly, chair was

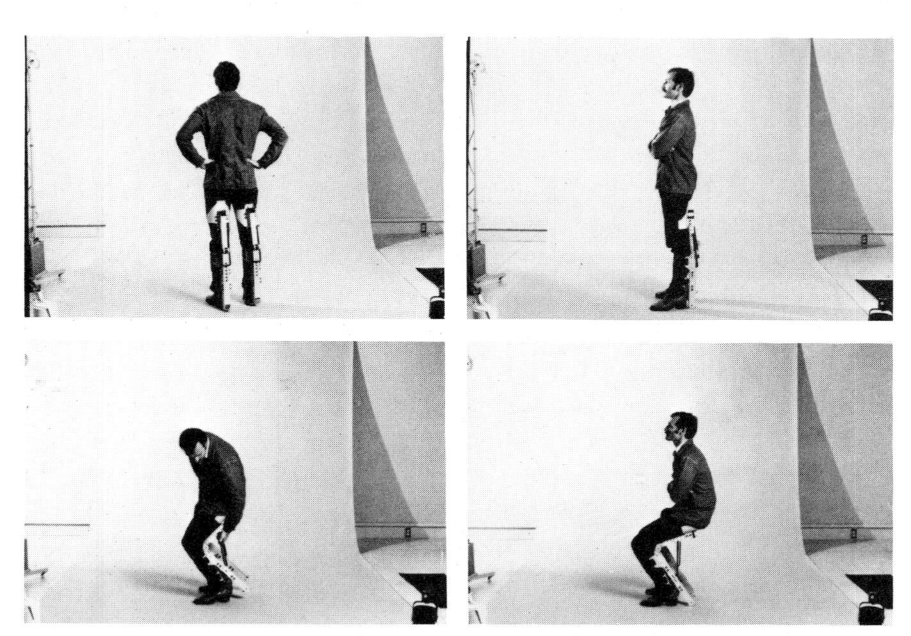

The chair is a prosthetic device for a cultural disorder called "Walking Upright." But rarely has the prosthetic aspect been so cleanly acknowledged as in the "Wearable Chair" designed by Darcy Robert Bonner, Jr. Strapped to the user's legs, it allows him to sit whenever he feels the urge.

popular as an army officer's campaign chair because of the ease with which it could be folded and moved and unfolded again. The same features made it popular as an American deck chair and as a prop for slapstick comedy.

According to Siegfried Giedion one of the most powerful elements of ruling taste in the nineteenth century was what he called "the furniture of the engineer"—the lounge that converted into a cradle, the bed that converted into a wardrobe, the chair that converted into a lounge, the table chair, the bed chair. Our space and housing shortages have revivified the trend, and it may be possible today to buy a side chair that opens up to sleep six.

In 1957 the Institute of Design of the Illinois Institute of Technology set out to determine the 100 greatest product designs. They went about it in a rather odd but highly symmetrical way, asking 100 designers, critics, writers and design teachers each to prepare his or her own list of the ten best designs. In the final top hundred there are fourteen chairs: the Thonet "Vienna" cafe chair of 1876, the 1903 director's chair, the 1928 Breuer chair, the 1928 Barcelona chair by Mies van der Rohe, the 1934 Aalto bent plywood chair, the 1938 Hardoy chair, the 1947 Eames plywood chair, the 1948 Saarinen "Womb" chair, the Finn Juhl chair of 1949, the 1949 Wegner round chair, the Eames fiberglass chair of 1951, the Brunswick classroom furniture of 1953, which included chairs, the 1957 Eames lounge chair, and the 1957 Saarinen pedestal chair.

Finding so many chairs in a list of the 100 greatest product designs raises the question we started with. Why does it seem to matter so much? What's so special about chairs? Well, remember that a chair is not an artifact of service but an artifact of culture. Not all cultures have chairs. And in cultures that have had them, not everyone has been able to afford them. But to people accustomed to chairs, their absence is a serious cultural deprivation, as the designers of prison and army barracks know.

Remember too that no other animal requires a prosthetic device for regular, ongoing use. A chair, after all, is a crutch for a condition that will not mend: walking upright. Designers like to speak of product evolution, but of course chairs have not evolved. *We* have. And once we came down from the trees or up from the sea or from wherever else, once we gave up the use of four legs for ambulatory purposes, we had to have a prosthetic device to rest on.

Or thought we had to have it. For while a chair is a crutch, it is not a crutch we need physiologically. And perhaps the chair's significance lies there: it represents the irreducible minimum of the unnecessary. A chair is the first thing you need when you don't really need anything, and is therefore a peculiarly compelling symbol of civilization. For it is civilization, not survival, that requires design.

A Pride of Camels

A camel is a horse designed by a committee. **Old Folk Saying**

A giraffe is a horse designed by a committee. **Alternate Old Folk Saying**

If two of us agree, one of us is redundant. **Bernard Benson**

Because design depends on individual talent, it is regarded by some as a solitary act. If it were an act, it might be. Certainly the richest design ideas, like the richest ideas of any other kind, come from single brains, rather than from cerebral executive committees or think tanks. And great designs, like great poems, paintings, and inventions, come from one fertile mind at a time. But the vision of the designer as the romantic loner of *The Fountainhead* is rarely valid or useful. Almost every design is in some respect collaborative. This does not mean that design is committee work. It does mean that before all but the simplest projects are given form, a designer has to depend on and work with a variety of people: managers, researchers, engineers, marketing people, social scientists and others. And this collaboration, this "working with," is design. Graphic designers will sometimes start by playing with forms to see where they lead, but the process does not normally begin at the drafting board.

It may begin in a meeting, unlikely as that has come to seem. The literary superiority of the King James Bible notwithstanding, group creative work is still presumed to be a contradiction in terms. The suspicion is sound enough; most committees operate like giraffes that have been designed by camels. But while design by committee represents a solemn commitment to mediocrity, design by collaboration is simply the way the world works, usually for the better.

The more complex a project, the more nearly inevitable that it be a group activity; and even very simple products involve a number of people before the design is complete. In the field of industrial design this sometimes gets twisted into an exaggerated notion of what designers themselves need to know: the "Renaissance Man" concept again. Descriptions of what one has to know in order to be an industrial designer are commonly so comprehensive and overblown that only Leonardo da Vinci (who has been—no kidding!—soberly cited by some industrial designers as the first in the busi ness) could qualify. The designer must fully understand technology, we are told. The designer must understand people and know their wants and needs. The designer must know how to organize space. The designer must know which materials are available at any given time and what each of them can do. The designer must be a master of fabrication techniques. The designer must be familiar with the workings and dimensions of the human body. The designer must be a master of marketing and sales strategies. The designer must be a planner, and an artist to boot. Also an ecologist. And so on, and so on.

Apart from all these musts and shoulds, what designers do

know and actually *are* able to perform varies considerably. The first industrial designers came from wherever they could be found: theatrical design, typography, engineering, architecture, commercial illustration. Design training has reflected that diversity. Depending on the school, a graduate industrial designer may understand something about engineering and hardly anything about art, or he may know a lot about art and know nothing about what goes on in factories; or he may know astonishingly little about either art or engineering.

Obviously, even the most earnest of industrial design spokesmen know, and when pressed will acknowledge, that the designer's versatility does not make him a master of any of the many fields his professional life touches. He is not so much twenty times a specialist as twenty times a layman, and that may work in his favor. The ignorance that a designer brings to a project is in itself a valuable commodity, and the history of industrial design abounds with examples of products that have been vastly improved by a designer's ignorance of the limitations of technology. Not knowing what *couldn't* be done, designers have drawn up plans for doing it. Some of these plans have been realized, but seldom by designers alone.

When a straightforward *product* is not the goal, the designer's role may be obscured, with the client valuing the designer's contribution but not knowing how to describe it. The designer is not always helpful in such cases, frequently describing himself as a "catalyst."

Whenever anyone calls himself a catalyst I figure that he must know even less chemistry than I do. A catalyst makes things happen, true; or it makes things happen at an accelerated rate. *But it does this without undergoing any change itself.* I don't trust inert design.

I also distrust the chemical analogy to creative action. Chemicals react. People respond. Unfortunately, too many people in too many meetings and institutions *do* react. That is what is wrong with them. What we need is not more catalysis, but more thinking and feeling responsiveness. But designers persist in seeing themselves as the Typhoid Marys of the communications process, carriers who remain unplagued by whatever it is they carry.

If design is collaborative, who are the collaborators? A map of the landscape of industrial design reveals a sort of Balkan territory bordered by engineering, sales, architecture, suppliers, product planners, competitive producers, advertisers, marketers, "human engineers."

In product design, the engineer is a principal collaborator. The first generation of American industrial designers were, as we have seen, used largely as face-lifters. As such, they had sales executives as clients. But during World War II, with consumer products replaced by defense products, designers were used more seriously because we take war more seriously. According to industrial design historian Arthur Pulos, "the designer learned to work for engineers, where before the war he had worked primarily for salesmen."

More important, the designer learned to work *with* engineers, although he did not always learn to work with them very well. Even at its most superficial, as "styling," design affects a great many products and kinds of products. The same designer may very well be responsible for a mechanical walnut cracker and an electrical circuit breaker, but what does "responsible" mean? Although designers may determine the final appearance of those objects, their general appearance—as well as their very existence—is determined by engineers. Or so it looks to engineers.

The designer-engineer relationship is shot through with misunderstanding, seeming conflict of interest, and just plain petty backbiting. Presumably the common interest calls for an intense collaboration, with each bringing his own expertness to the problem. But to say just that is like singing "The farmer and the cowman should be friends." Rodgers and Hammerstein were right about that, but the end of the lyric was not the end of the conflict.

True, the goals of engineering and industrial design are not, in any sane situation, in conflict. But industrial situations are no saner than any other, and happy endings are harder to achieve in product development than in musical comedy. The designer and the engineer *are* friends, at least working friends, whenever they get through the mythology that divides them. It isn't easy.

According to the myth prevailing among engineers, the designer is all gloss, a stylist equipped by temperament and training to do nothing more than frost the bread of life so it can be sold as cake. He is the cowboy, dashing on a palomino, dazzling with rope tricks, addicted to boasting, showing off, and buying tax-deductible drinks for everybody. His guitar attracts the attention of management, but his high-heeled boots are impractical for walking on solid ground.

Well, that *is* true of some industrial designers and used to be true of a lot more of them. But when he confronts the engineer, the designer tends to be damned if he does and damned if he doesn't. If his contribution is merely styling, he comes across as a parasite living off the organic health of engineering. If, however, he contributes to structure and to function, he is accused of stealing the professional engineer's birthright.

On the other hand, consider the mythical engineer as the industrial designer sees him. Narrow and uncreative, he is a highly trained mechanic, congenitally unable to view a product in any context larger than the working of its parts. He is wholly oblivious to the requirements of marketing, use, convenience, and appearance. He is the farmer: solid, stable, and square; but so set on plowing a deep straight furrow that he can't see the world on either side. Well, some engineers are like that. But the engineer is also damned if he does or doesn't. Industrial designers begin to resent him the moment he becomes concerned with the very things they complain that he doesn't care about.

The plain truth is that, even if appearance were the only criterion, engineers have a better record than designers have. They have created more beauty, if not less ugliness. As art critic Sheldon Cheney observed, "the machinery in the power house has a potent line-and-form fascination that anyone alive to art must feel," and certainly some of the most stunning examples of twentieth century design are technological products that have had no design treatment as such. Although the beauty in the multicolored printed circuitry of electronic data processing may seem esthetically arbitrary, dictated by the restrictions of function and materials and

Oil refinery photographed by Paul Strand.

cost, a variety of forms are possible within those restrictions. Choices had to be made, and some sensitive engineers made them. And those very restrictions—function, materials, tooling, cost—are the restrictions within which any designer makes his choices.

Constraints are part of the context of any responsible design.

Crediting is a sensitive issue in the designer-engineer relationship. That credits ever became an issue is pretty much the fault of industrial designers whose orientation was toward art, and underscores the extreme difficulty of ever understanding precisely what it means to say that something has been "designed."

"I have just redesigned my airplane," an industrial designer told me. "Inside and out." Actually he *had* designed it inside and out, to the extent that industrial designers design airplanes at all; but nothing he had done to it affected the most important attribute of an airplane: its ability to fly. As both pilot and designer, he understands this better than I do, but the descriptive problem remains.

On the other hand, the designer of a kettle has usually done just what is implied: designed the kettle. The difference is one of complexity and control.

Don McFarland, who is trained both as an aeronautical engineer and an industrial designer, describes an interesting difference in the two operating styles. An engineer, according to McFarland, customarily works from the inside out. That is, he is trained to solve problems by thinking first in terms of technical details. But the industrial designer normally works from the outside in. His thinking starts with the complete product as it would be used by someone, and works backward into the details required to make the concept work. To an engineer this may seem like beginning at the end, and in a way it is. Paradoxically, this backward way of working explains why it is important for the designer to be involved in a project at the very beginning: if he is called in after the details have been worked out, there is not enough room for the development of a concept responsible to the user's needs.

The two very different approaches have to do with two very different roles. They can be usefully oversimplified in the following way. Let us suppose that the product under development is an electric fan. The engineer is charged with seeing to it that the fan will work. The industrial designer is charged with seeing to it that someone will—and conveniently can—work the fan. This represents a world of difference in emphasis. The engineer's main concerns have to do with the interaction of components and materials. His job is to develop the best product consistent with the state of technology and the anticipated market price. The designer's job is to make sure that the product is one that will be bought and used—not, one hopes, in that order of importance, but usually in that sequence.

Yet to put it that way makes it sound as though these are discrete steps in a creative assembly line—with the engineer completing phase one, then turning the job over to the designer for polishing. Design in such a case would be simply laid on, "applied art" in the worst sense. A lot of products do get "designed" in this wasteful way, but a product, like a person, can have no integrity as a split personality, and product integrity generally requires that engineer and designer work with each other, rather than take turns. Design is done within constraints and engineers know what the technical constraints are. (Unfortunately they may see constraints under every bed.)

Some of the sharpest insights into industrial design have been supplied by people who are not designers. Of these none have illuminated the process with more direct brilliance than J. Bronowski, when he described design as taking place within a triangle of forces: the tools and processes by which an object is made, the materials of which it is made, and the end use. "If the designer has any freedom," Bronowski said, "it is within this triangle of forces or constraints."

There are various approaches even within that triangle, depending on the talent of the designer and the scope of the project. If the product is a toaster, the design response may take a number of different forms.

1. Styling. This is essentially packaging, and the designer's role is to house the heating mechanism in a more appealing case.
2. Another designer may go further, rearranging the mechanism to increase heating efficiency or reduce size.
3. A third designer may begin with the heating mechanism and redesign *it*.

4. Still another might (and did) study end use and, noticing that for many Americans white sliced bread has become the comestible of last resort, design a toaster that will accept bagels, bialys, scones, *pain au chocolat.*
5. Another designer, realizing that virtually every household that buys a toaster can already make toast in the oven, might devise something to make oven toasting as convenient as pop-up toasting.
6. An even more rebellious designer might work on a way to treat bread chemically so that it toasts itself upon being unwrapped.

Those approaches are arranged in order of decreasing probability. The first, alas, is what most designers would be expected to do for most clients. But it is worth noting that all approaches depend upon a designer's learning what engineers are already supposed to know.

Here again, the designer comes equipped with what may appear to engineers as technical ignorance. In many cases this is not a matter of appearance; it *is* ignorance, but ignorance of a kind that can helpfully supplement the technical knowledge and specialized skills of the engineer. It is neither necessary nor desirable that all members of a manufacturing team know the same things. A little technical misunderstanding can do a power of good.

Early in his filmmaking career, designer Saul Bass was searching for a thematic image to use in the title sequence for a movie called *Carmen Jones.* When he saw a still photograph of an undulating flame, he knew the effect he wanted: to animate the flame. He made some sketches and described the effect to the studio engineers, who said it could not be achieved photographically. Bass insisted that it could be and had been. The engineers proved that it couldn't be done, by trying to do it and failing. Bass promised to prove it could be done by producing the photograph that had given him the idea in the first place. The trouble was, he couldn't remember where he had seen it. He ransacked his library, going through volume after volume of picture books, and finally found in triumph the picture he was looking for. The triumph faded: it was not a photograph of flame at all, but of a jet of water coming from a bathroom tap. He had looked at it upside down!

The following morning a shamefaced designer went to the studio to apologize to the engineers. Before he had a chance to begin, the chief engineer said triumphantly, "Well, we did it." Operating from Bass's false premise, the engineers had invented a technique for making it valid.

A more nearly typical experience is that of a designer who was asked to redesign the housing for a stereo speaker. Some of the changes he wished to make would have required corresponding changes in the guts of the product. Engineers objected. The dialogue went like this:

> "You can't move that. It has to be where it is."
>
> "What if you did move it?" the designer asked.
>
> "Well, if you did move it," was the reply, "it wouldn't work. Not unless you moved this too."
>
> "And if you did?"
>
> "You *couldn't*. Oh, I guess you could," the engineer said, "but that would mean . . ."

In the end the engineers found that they could accommodate the designer. That in itself is a terribly unimportant objective. But their exhilarating discovery was that the necessary changes led to superior sound reproduction.

Depending on the project, the models for interdisciplinary design vary a good deal. In the case of appliances and other consumer goods, engineer and designer may work so closely together that afterwards neither will be able to say clearly who did what. In the case of a computer system, however, the mathematician responsible for the information theory and the engineers responsible for translating this into working electronics have roles that are very far removed from that of the designer responsible for organization of the console and color of the panels.

The object of design and engineering attention is not invariably a product, for in both engineering and design the emphasis on product has often given way to an emphasis on system. But the two disciplines differ in approaching systems too. The engineer must see a system as a series of products and components that can be related to each other in various configurations. The designer is concerned with those possible configurations, but he is also concerned with the related systems of graphics, packaging, marketing, and above all, operation by users and *systems* of users.

Systems engineering is a legitimate specialization. And because a "systems approach" reflects the part of our culture that began with Henry Ford, it has found its way into design and manufacturing generally. Or at least it has found its way into the jargon of design and manufacturing. "System" sounds more complicated and demanding than "product," for it suggests that a multitude of diverse elements have to be coordinated into a single whole. An advertisement in a newspaper sports section describes a rain hat as "not just a hat, but a headwear system." A New York department store confides of its mattresses, "We like to call them sleep systems." An interior designer describes the layout of space in an office building: "It isn't an internal design problem at all—it is a mini-urban systems problem."

In describing man, the indefatigable Bucky Fuller speaks of "a self-balancing, 28-jointed adaptor-based biped; an electro-chemical reduction plant, integral with segregated stowages of special energy extracts in storage batteries, for subsequent actuation of thousands of hydraulic and pneumatic pumps, with motors attached; 62,000 miles of capillaries. . . ."

And yet there are those who love us, perhaps because we are the only systems that are simultaneously open and closed. No man is a wholly owned subsidiary. Each man is a partnership unto himself.

The collaboration between engineer and designer permeates the design of three-dimensional products, of hard goods or hardware. Other collaborations in design are no less intense, although the cast of characters changes with the product. The design of an exhibition may require the joint effort of designer and physicist, designer and historian, designer and sociologist, designer and ecologist—depending on whether the exhibition deals with Galileo, Keynesian economics, Maltese armor, or predator-prey relationships. *Producing* the exhibition may require an equally intense active collaboration with sound engineers, material experts, photographers, and writers. Even the styling of products is not an individual matter. While the goal may be superficial, the means of reaching it are not. Here, designers join with sales and marketing people and with marketing research psychologists.

Getting into bed with so many people can be an enriching experience, but it is not risk-free. Hermits catch very few communicable diseases. One of the penalties that designers must pay for collaboration with social scientists and business executives is the jargonization of discourse. There is even jargon to describe the process by which jargon is produced—as in *interface,* which has become a verb ("Let's interface"), a gerund ("We were just interfacing"), and a general stumbling block to clear communication ("You can't interface with someone you don't access").

In space planning, interior design, and exhibitions, the industrial designer's collaborator is frequently an architect. For that matter, the industrial designer may himself be an architect. Eliot Noyes, George Nelson, Charles Eames all began as architects, as European industrial designers commonly did and often do. Henry Dreyfuss went so far as to call industrial design "quasi-architecture," but then he also went so far as to call shaving "minor surgery."

Industrial designers are often well suited by experience and approach to the design of buildings that "do things." They boast, with some validity, that they have more interest and experience in relating people to working space than most architects have. Where the major architectural requirements are merchandising, consumer appeal, corporate identity, and work efficiency, industrial designers may function as architects. They design shopping centers, supermarkets, gas stations and factories, and do space planning of office buildings.

According to Brooks Stevens, a business building is a building designed to make money. William Snaith, who designed a number of them, made the same point when he said that "a shoe store is not a wall structure, but a mechanism for selling shoes," and that the industrial designer is a designer of mechanisms.

The association of architecture and industrial design has not been hailed universally, as Sibyl Moholy-Nagy complained:

> The identification of industrial and building design gained ground rapidly because it provided untalented and frustrated architects with mechanical standards and book-length theories: it cut costs for expanding shelter needs and it gave to industrial design an intellectual pedigree, derived from architects turned into designers. Each show of automobiles, industrial containers, home furnishings and posters at the Museum of Modern Art was a coming-out party presenting a newcomer in the field of design to its elders. But the forced alliance did neither of them any good. Architecture started to compete with product design for visual sensations and "progressive use" of synthetic materials, forgetting that an out-dated appliance disappears in the scrap pile while a shoddy building keeps on standing . . . industrial design borrowed from architecture structural shapes, predicated on space and closure, producing extravagances which are senseless without the space function. . . .
>
> Only by a proud uncompromising divorce of this art from industrial design as the shaping of expendable objects, by a separation of permanence from obsolescence patterns, of idea from expediency, can twentieth century man save himself from turning into a gadget himself, packaged in mass-produced containers and disposable like them.

The dangers to the twentieth century human personality are grave indeed, but it is hard to seriously consider the alliance of architecture and industrial design a major threat. The prospect of mass-produced containers is not all that bad either for a nation that is now poorly housed in ways that Roosevelt's famous indictment never anticipated. Unfortunately, the most enthusiastic assessment of the mass-produced house is still to be found in its advance notices.

For half a century industrial designers have prophesied the

well-designed, affordable manufactured house, unveiled plans for it and reported it as waiting in the wings. Yet for a long time the closest approximation was the mobile home planted immovably in a trailer court, for building codes effectively blocked the appearance of a real manufactured house. Today we have modular houses and modular parts of houses but no one hails these as design triumphs. What we need is a sort of Volkswagen of housing. Current economics and conventional buildings militate against its development, but are no technological reasons not to have one.

The collaborative pattern of design admits exceptions. Some graphic designers insist on working alone, and necessarily restrict thereby the scale of projects they take on. But in three-dimensional design, as specialization increases the individual designer is increasingly supplanted by a design *team.*

Almost every design project requires two collaborators—one at each end of the project—whose participation is frequently not identified as participation at all. The first of these—and generally the first of all collaborators—is the client. Technically, and legally, the client is the person or corporation that hires the designer. The working relationship between them affects the designs we get. And while most readers are neither clients nor prospective clients for design services, some of you are, so we ought to look briefly at what clienthood involves.

A client's minimal responsibility is to hire a designer. For many, even this is extremely difficult. I have seen clients who can with ease locate competent plumbers, doctors, and gardeners, but who are utterly confounded by the thought of getting a designer. They have no confidence in their ability to assess the designer's work, and no clear understanding of what that work should be. How do they know they won't be taken? How will they even know if they *are* taken? Do they start with the Yellow Pages? Do they ask a friend?

The answers depend on so many variables and possible combinations of variables that no satisfactory answer can be given apart from the specific problem. But here are a few tips on choosing a designer.

Certainly ask your friends. Ideally these will be friends who have some professional basis for judgment, but as in anything else, any referral is better than a cold call. You will at least find out something about a designer's approach, his ability to understand someone else's problem, his rates, his reliability.

Look very hard at the designer's work. Ask questions about it. Don't rely on general impressions alone, but try to find out what the problem was and why the particular design represents a solution to it. But although your general impression is not infallible, in the end the decision rests on personal judgment, and there is nowhere else it *can* rest. The best advice in this regard is paradoxical. You are cautioned against assuming that effective design decisions are simply based on what you like; but in choosing a designer there is no intelligent way to ignore what you like and no reason why you should want to. Unless you are a masochist you will naturally look for work that pleases you. Just as naturally you will then find out who did it and whether or not the client was satisfied.

Consider two or three designers. You need some basis for comparison.

Remember that the designer is probably not a specialist in your business. Remember too that you don't want him to be. That's not what he's for. Perhaps the most common error in designer selection is to select the designer for a sewing machine on the basis of a sewing machine he has designed for somebody else. Conflict of interest would keep him from taking you on as a client if he were still associated with a competing company, but specialized experience is the wrong kind of thing to look for in any case. The most important thing to check out is the designer's thinking, not his history. Try instead to discover whether the designer's general approach is compatible with your own, and whether he brings to his designs the kind of imaginative thinking you want in your own products. If it is genuinely imaginative thinking, there is no way to know in advance what you will get. A good design office will try to minimize your surprise every step of the way, but a design that does not surprise you at all is as dull as life would be if *it* didn't surprise you at all.

Once you have a designer, what do you do with him? Or, if you are a designer, what do you do with a client? The first thing to do is trust him. I don't mean to suggest that all designers are trustworthy; they are not. I do mean to suggest that an effective design relationship is based on trust. And it can't just be honeymoon trust; it has to survive first love. Some clients suffer from a malady one might call the Groucho Marx syndrome, in honor of Marx's legendary refusal to join any club that would have him as a member. I have known clients to search for the one "right" designer, only to discover that his feet were clay upon arrival. In most cases there probably is no single "right" designer. But a client-designer relationship, if nurtured, can over a period of time seem to have been made in heaven.

Most designers acknowledge that the client must be "involved." But involved how? Where? At what point? There is a good deal of ambivalence about this. On the one hand, designers want client involvement and know they need it. On the other hand they are afraid the client may interpret involvement to mean the arbitrary imposition of his personal design preference—e.g., "I don't like green."

That is a real risk, but in my opinion the more genuine the involvement, the less of a risk it is. Perhaps the proper question is not "Who is the right designer?" or "How much client involvement should be encouraged?" but "What roles do client and designer play and how do they play them?"

The idea of a role suggests that the client is active, not passive; and temporary, not full-time. This is not unprecedented in business life, where corporate responsibility is figuratively segmented along the lines of which "hat" an executive thinks he is wearing at any given time. The ultimate corporate schizophrenia is expressed in such executive assertions as "With my marketing hat on I like it, but when I put my advertising hat on . . ." but the metaphor has something to teach us about corporate roles. There are marketing hats and advertising hats and production hats and also party hats. Why are there party hats? Why does anyone, after a couple of drinks, want to put on a fireman's hat or a policeman's hat or a lampshade? Why do politicians, even in the 1980s, continue to campaign in hardhats or Indian headdress or yarmulkes?

Well, a hat lets you assume a role with minimal commitment. You get *into* a jacket, but you put *on* a hat; and can take it off with ease if the going gets tough. The danger is that a client will put on a hat, in effect, and keep his commitment low. Being a client is, after all, a part-time activity. If you meet someone at a cocktail party and ask, "What do you do?" he never replies "I'm a client." He is only a client when he is clienting. Because the designer is full-time and professional it is his responsibility to encourage client involvement.

This frequently means trouble, but what meaningful relationship does not? The only trouble-free client-designer relationships are those in which the client's role is simply to foot the bill, and they are only as trouble-free as the bill is. In such a case, the client is really just a sponsor, having possession of the property but no responsible ownership of it. Like sponsors generally, such clients exercise their rights in the matter through arbitrary censorship, if at all.

That sort of relationship is entered into at great cost. If the client's role consists merely of bankrolling the project, the designer's role is that of an exotic menial. He is menial because his services are required for low-level objectives, to be considered only after the *real* business decisions are made, and exotic because no one really understands what he does. Although this is a horrendous misuse of the designer and of the design process, it is always done with the designer's collusion. To accept this role is to agree implicitly not to be involved in matters of substance, but simply to make things nice.

But making things nice is not making things right. And it is in the rightness of things that consumers have a stake. More than a stake, a role to play. For the designer's final collaborator is the end user.

To see the end user as a collaborator is to think of the design process as continuing beyond production. And, for better or for worse, it does. There is an implicit contractual relationship between designer and user and—as with other contractual relationships—the contract may be betrayed.

We have already seen some ways in which the designer can betray the user: chairs that don't support the body, instrument meters that are hard to read, connections that don't hold together, appliances that can't be cleaned, literature printed in illegible type. Ironically, excellent appearance design can be an instrument of betrayal, as in the Italian typewriter that makes you feel like a war correspondent but makes the simplest letter look as if it were typed at the front, the German line of appliances that are lovelier than ours but don't heat, cook or mix as well as their inelegant counterparts.

A rarely identified but major U.S. health problem is iatrogenic disease—the various ailments and illnesses brought on by the practice of medicine itself: by the careless nurse, the ignorant doctor, the disorganized hospital, the misread x-ray, the side effects of drugs (often wrongly prescribed in the first place) and therapeutic machinery. The cure in such cases is not only worse than the disease; it causes the disease. Adverse drug effects are one of the chief causes of hospitalization, and one out of every five hospital patients acquires an iatrogenic disorder. Modern medicine can hurt more than it helps.

So can design; and we are heir to a great body of iatrogenic design disorders—products that have been made worse by the designer's touch. This morning's mail brought a brochure printed in brown ink on brown paper with a black sketch accompanied by handwritten notations laid over the type for decorative effect. It was almost unreadable, yet the copy must have been perfectly legible when typed. A leading industrial designer admonished his disciples to "never leave well enough alone." It is a good slogan but bad medicine.

Why are iatrogenic design disorders so widespread? Because some designers, like some doctors, are inept, stupid, or greedy; because all designers, like all doctors, make some mistakes; because many consumers, like many patients, demand relief they can't get from ailments they don't have. And, as in medicine, new therapies have to be invented to treat iatrogenic disorders.

Design betrayals occur in such abundance that a number of auxiliary industries, in the words of Elizabeth Gundrey, "thrive on the dropped stitches of incompetent designers." Ms. Gundrey, a leading consumer goods critic in England, devoted an article to the subject in the British magazine *Design.* It is a witty and perceptive argument, loaded with facts we already know but may never have thought of before in this context. I know I didn't. Her examples are endless, as yours and mine would be too. Here she is on the subject of sinks:

> Because the grids at the drain-hole let through bits which will later clog the pipes, little cages are sold to stop this happening. Because the flow from a tap often strikes the sink at an angle and speed which cause splashes, rubber anti-splash devices are sold . . . . Because the recess in the sink rim, intended for soap, does not drain but just collects a lot of horrid slime, ingenious minds have worked out innumerable gadgets—spiked, magnetic, ridged—to keep the soap out of its own puddle.

Ms. Gundrey's examples don't stop at the kitchen. She cites the load-spreading plastic cups designed to keep the feet or casters on furniture from digging into the carpet. The nonslip mats or special bars for bathtubs. The heel-grips to keep shoes on. And she points out that "so inadequately does the telephone perform its function that there is a brisk trade in special dials . . . shoulder rests for the handset . . . and wiggly spirals to stop the flex kinking. Not one of these would be needed if the telephone had been adequately designed in the first place."

Not that the corrective devices are always left to auxiliary industries. Ms. Gundrey notes that "some manufacturers spot for themselves the deficiencies in their own designs, and start selling their fumdubulators to add." Vacuum cleaners that don't suck up hairs properly can be ordered with special attachments that will. As for cars, she reminds us that "never was there any mass-produced product which demanded the immediate collection of so many barnacle-like extras." For confirmation, which none of us need, any auto accessory shop has a huge inventory of products that exist only because the auto industry has failed to design adequate steering wheels, windows and seats.

But betrayal works both ways. I once walked down Fifth Avenue in Manhattan admiring the woman walking ahead of me—a willowy long-haired brunette in a gray pantsuit, carrying a leather shoulder bag. From the rear the effect was striking, but when she turned and stopped at a shop window I discovered that she was chewing gum with her mouth open! She had a perfect right to do that, I guess; and perhaps I had no right to feel betrayed. But I am not so sure about that. After all, she had designed an effect, with the help of Yves St. Laurent and Ferragamo for all I know, and was under some obligation to it, if not to me or to the designers.

That kind of betrayal happens all the time. The entrance to a building is tailored to the teeth, with every detail planned for elegance. But its most conspicuous visual feature, the one that sticks in the visitor's mind, is not anything the designer provided, but a hand-scrawled note on a shirt cardboard that reads "Please close door after entering."

There are, at least, two kinds of betrayal implicit in that situation. The client has betrayed the designer by introducing an impromptu design or anti-design of his own. But, chances are, he has done this because the designer has betrayed *him* by not designing the door to close by itself in the first place. Design always has some communicative aspect to it and therefore always depends for its efficacy on behavior conducted in good faith. The relationship between designer and end user is the basis of humane design.

Suitable for Framing

People can insulate themselves from painting, sculpture, literature, music, the dance. But they can't avoid exposure to the design of buildings and green spaces. Since urban design, architecture and landscape design are the only arts that cannot be avoided, they have a uniquely public character. Martin Myerson

boss I am disappointed in
some of your readers they
are always asking how does
archy work the shift so as to get
a new line or how does archy do
this or do that they are always
interested in technical details
when the main question is
whether the stuff is literature
or not Don Marquis, archy & mehitabel

Sometimes the main question is whether the stuff is art or not. If the industrial designer is the artist in industry, then industrial design must in some sense be art. But in *what* sense?

The high school I went to offered a course called Commercial Art. Students who drew well, as I did not, took the class, as I did not. So far as I know no one from my home town ever became a commercial artist. In fact hardly anyone from anywhere seems to have become one. I know a lot of people who call themselves artists, but very few who call themselves commercial artists, although an advertisement for an institute of culinary arts speaks grandly of its field of study as "the art that pays." Jacques Barzun writes that people in the teaching profession will go to any length to keep from being called teachers. Well, people who practice commercial art will go to any length to keep from being called commercial artists. To call oneself an artist is pretentious. To call oneself a commercial artist tempers the pretension with self-disparagement.

This is not a matter of snobbery entirely. In part, it reflects our current specialization of activities: commercial artists include illustrators, keyliners, layout people, sign painters, renderers, photographers, graphic designers and other specialists. "Commercial art" implies a distinction that is difficult to make, but it wasn't always. The term was once quite simply synonymous with *applied* art. Art addressed to nonartistic concerns. Art that was art only in respect to the ingredients used, not in respect to what it aspired to or accomplished for the human spirit.

In an age when all distinctions are blurred, it is especially important that those we can still make be valid. It is harder than ever for the eye to distinguish between art for the enrichment of souls and art for the enrichment of capital investment. Is there a point at which they are interchangeable? The balloon that had begun to swell in the twenties with Dada finally popped in the sixties with the comic strip balloons of pop art. Galleries such as Greengrass in New York and O'Grady in Chicago specialize in commercial art; many other galleries across the country handle it. The Louvre has mounted major shows of the works of Charles Eames and Push Pin Studios, which was for a while a sort of Black Mountain College of

Poster for The Threepenny Opera designed by Paul Davis.

Commercial Art. Saul Steinberg's magazine covers are as at home in the Whitney Museum as any of his other paintings. The work of Corita, Paul Davis, Tomi Ungerer moves from ad agency to gallery show. (Corita not only contributes to advertisements, but finds parables in them that she works into her art.) The advertising graphics of Lou Dorfsman at CBS have been featured at Tel Aviv's Museum of Modern Art. *Punch* and *The New Yorker* illustrations by Andre Francois appear in museums, and New York's Museum of Modern Art in 1976 mounted a popular show of the work of Milton Glaser. If Andy Warhol had not been invented he would have had to exist.

To talk about taste moving up and down, or even from left to right, is to pretend that we are observing it objectively, as if it were a chair or a flying saucer. Yet we still speak of it as a vertical concept, a commodity to be raised by the availability of Knoll chairs or Rosenthal china, or lowered by the manufacture of dinettes. In this view public taste (which everyone but the public is concerned about) can be escalated or de-escalated as mercurially as a war in another country. It is tidy to suppose that taste can be sent up on a dumbwaiter or elevator, for this suggests that we need only know which rope to tug, which buttons to push. We don't have such buttons, if we had we probably wouldn't know which ones to push, and fortunately it doesn't make all that much difference.

Industrial designers, more tasters than tastemakers, built much of their success upon a gift of timing as tangible as Jack Benny's. The successful designer kept his clients at work on styles that were just a trifle ahead of what the public would accept. In order to do this, he had to have on his drawing board designs that were far more advanced than the public would accept, but that represented saleable concepts for the future, when the public would catch up, as designers always predict it will. Public taste, in other words, is continually rising, although of course by the time it gets to where you and I are, we have soared higher still: taste never catches up with tastemakers.

Curiously, while the vertical view implies that design is ahead of the public on some scale of enlightenment, it fails to recognize this as a scale that designers themselves have either devised or helped devise.

The public is not unready to accept the superior; it is unready to accept the unfamiliar. To the extent that designers consciously lead public taste, they do it by introducing carefully measured doses of the unfamiliar on the grounds that if the dose is too small the public won't notice it and if it is too large the public won't swallow it. But although designers in this sense lead public taste, they do not lead it to any *destination.* It is fallacious to suppose that such leadership represents an advance. All it represents is a change. The mini advances toward the maxi, the stark advances toward the ornate, the formal advances toward the casual. With all these advances one would expect an occasional retreat. But—as in the

United States Marine Corps—there are no retreats: there are only advances in the opposite direction.

The New Yorker once carried an advertisement for Tiffany, headed "Let's Talk About Taste." The ad testily scolded the innocents who think that taste has something to do with what they like: "They confuse their personal preferences with taste. . . .They say 'I like this' or 'I don't like that,' not realizing that this has nothing whatsoever to do with taste."

If taste has nothing whatsoever to do with what one likes, what on earth does it have to do with? The ad told us: Tiffany's own taste is "based on good esthetic principles." It did not say whose.

Oh, we all know something of what the copywriter probably meant. He meant that taste can be cultivated, even "improved." (I can always tell when someone's taste is improving: it moves closer to my own. But the ad is reminiscent of the character in the old Krazy Kat strip who kept saying, "Even if it was good I wouldn't like it.") It's fair enough, although hardly in the vanguard of esthetic revelation, to suggest that there is more to taste than just saying "I know what I like." But to suggest that there is *nothing* more to it than learning some "esthetic principles"—as if these, like Strontium 90, were discovered wholly apart from what the discoverer liked—is worse than the fallacy it seeks to correct. For, in addition to being absurdly oversimplified, it is snobbish in a way that is forever suspect. But profitable. For although Tiffany seemed to be cracking the whip on "personal preference" (a phrase that the ad used as if it were synonymous with ignorance), the advertisement mellows to a finish with a tidy way out: "At Tiffany's even the youngest bride can use her own personal preferences to her heart's content, because she can be sure, no matter what she selects, it will be both good quality and good style." This lets any paying customer have the pound cake of preference while eating the *Galette Strasbourgeoise* of certified good breeding.

In 1962 a magazine called *Environment* was launched, although not kept afloat for long. One of the first of the many American magazines to take that word as all or part of their title, it may have looked like an early entry in the ecology movement, but according

to its publisher, was a forum in which "the architect, interior designer, landscape architect, painter and sculptor could coordinate and integrate their efforts to achieve the optimum in home environment, thereby reinforcing our most basic cultural unit."

"Most basic" though it may be, the home environment was not so basic as to fall into the domain of the individual inhabitant. *Environment* had a mandate to "fight the do-it-yourself trend among homeowners," on the grounds that specialists are necessary for the avoidance of esthetic and other mistakes. Beauty, in other words, is in the eye of the paid beholder, and designers are mercenaries threatened by the patriotic amateur who will fight for no pay. Prostitutes do not as a rule resent other prostitutes, for they understand free competition and accept it; what threatens prostitution is promiscuity.

Also in the sixties a group of well-meaning New York society matrons formed something called the American Institute of Approval, Inc.—a nonprofit consulting organization that offered two services. For a fee of $1,500 a day the ladies would tell manufacturers whether or not their products were tasteful. If they were, the products would be awarded certificates of approval—like the one that *Good Housekeeping* assigns to advertisers. The Institute was made up of ladies who—in their own words—"have won for themselves, or been gifted with, superb and widely recognized good taste." It was supposed to "serve as a basic criterion for American taste in the best sense of the word," and its founders took neither themselves nor their mission lightly. "Never before," they proclaimed, "has a group of similar prestige come together for the benefit of the commercial world and the public good." The ladies were quick to point out that they were not designers but "decision makers between alternate products and designs" and they recommended that their taste could most profitably be tapped "after the commercial designer has presented his designs and at the moment when top management must make its decision."

At first I thought this was a gag, a Helen Hokinson idea for a musical starring Beatrice Lillie. But the ladies were in dead earnest and some designers took them seriously as a threat, if not as a public service. Peter Muller-Munk, probably the most urbane of American industrial designers, complained to me, "Now someone is going to tell my clients whether or not my designs are in good taste." At least one designer welcomed them: the design director for a major appliance manufacturer urged his management to submit his designs for approval, on the grounds that he had nothing to fear from a qualified judgment.

The Institute was not a wholly new phenomenon. Designers have always been plagued by the number of design decisions made by the client's wife. What the Institute of Approval had done was to institutionalize the client's wife, in an effort to provide for designers and manufacturers the same taste guarantee that Tiffany was providing for "even the youngest bride."

There have been serious efforts to support taste institutionally. England's highly respected Council of Industrial Design maintains a design index and permanent showroom in London, set up initially to assist manufacturers in designing profitable products, but charged as well with elevating public taste. Periodically, comparable agencies are urged on Washington, but few people in or out of government seem comfortable with the idea.

Taste is not stable enough to be institutionally approved. I don't mean simply that someone may have exquisite taste in furniture and dreadful taste in architecture, although that is certainly possible. I mean that one can and does have several tastes, any one of which may direct his choices in a given area at a given time. We speak of the conflict between champagne taste and beer budget, but one can be fond of both beverages, depending on the occasion and time of day. The diversity of options available today makes this far more likely than it used to be. Many of the same people watch "Wide World of Sports" and "Masterpiece Theatre," disturbing as this may be to some masterpiece lovers. (Boxing fans are less disturbed.)

I was in college in 1949 when Russell Lynes's *The Tastemakers* came out. Lynes pretty well laid the subject to rest then, but we won't let it rest easy. Lynes probably knew this, since his book's parting advice to us all was to stop worrying about taste since no one could define it anyway.

Our confusion about taste is nourished by confusion about whether design is art. Some designers view the artistic component of their profession as separate but equal. To design a chair or a washing machine or a spectrophotometer or a cereal box or an annual report is, they hold, as much an artistic endeavor as to create on canvas with paint or in space with clay. Museums have been known to encourage this view, and it is a tempting one.

At the extreme, the temptation takes the form of claiming that commercial art is *superior* to fine art, just as certain southern communities used to argue that, in their separate but equal school systems, blacks really had the better facilities. It is true that, although the motives of designers may be sharply different from the motives of fine artists, the designer's ability, talent and development often *are* superior; and the identical skills and tools are called for.

But, as the Supreme Court has taught us, separate can never be equal. Form and content are inseparable in art, whereas in design the content usually antedates the form. A painting is an answer to a need that cannot be conceived apart from the painting that answers it. Design on the other hand, is an answer to a need that can be discussed independently, and that could be answered by a number of alternate solutions.

New forms," says Milton Glaser, "cannot be generated in the area of design without great difficulty . . . the painter takes on the production of the difficult and tries to create a new language . . . even a poor painter is involved in that. But applied art can never work in the new because it must always convey information. It always works within what is known and introduces the new as an element in that."

Hilton Kramer explains: "The designer must hand over what he has invoked to the hazards of contingency at the very moment that the artist will want to transform it into a state of meaning. Because the designer wishes to affect our lives more directly, he speaks *for* our experience only remotely if at all. Art illuminates experience without asking us to become something other than what we are. The motive of all design is to change lives . . . even the most prosaic design always aspires to an ideal, while art addresses itself to actuality."

The ideal to which design aspires may itself be prosaic. George Nelson writes: "It is possible that on some abstract scale of values . . . a good door knob is better than a bad crucifixion. However, between the door knob and the painting, there are very real differences. Despite the wisdom and profundity of even the greatest designer, a door knob is, after all, still only a device to open a door. No matter how beautifully designed, how functionally appropriate, it is most unlikely that this object will arouse more than a very limited number of emotions."

Of course nothing is as simple as it seems, not even a door knob. It really is not "only" a device to open doors, for the design affects the quality of experience. A door knob may be a device to open doors elegantly, or to open elegant doors. (The standard American folk description of inappropriate detail is "a golden door knob on a shithouse door.") Moreover the door knob is not isolated (nor is the painting, although we are more likely to treat it as though it were). It is attached to a door, and relates to other objects in the room as the door opens. As objects go, the knob is not as insignificant as it looks at first glance. It is, after all, what you grab when performing two of life's most poignant acts—coming and going. How many scenes in drama and in life reach their peak when one party lays a hand on the door knob and the other says, "Wait!"? But does the *kind* of door knob being grabbed make any difference?

Certain critics argue that motivation is irrelevant in understanding art. It is not irrelevant in understanding design. Why does a designer create a door knob? He may have a client in the door knob business. Or a client who isn't in the door knob business, but thinks he may as well be. A nobler, or at least nicer, motive is that of the architect who discovers that there is no stock hardware available that is right for his doors. The door knobs in the Seagram Building were designed by Philip Johnson for that reason.

The difference between art and design becomes apparent in the way artists and designers go about shaping their talents to a cause. The designer, early in the game, asks: is there a market? What kind of market? What kind of people would use a product like this? How many of them are there? How much are they probably willing to pay for it? Where do they live? Would they be willing to polish it? Are competitive products available to them? How good are they and how much do *they* cost?

Those are all genuine considerations in making things for people. The painter makes things for people too, but he makes them for himself first. In theory a painter doesn't ask what's selling, although as a matter of record, a great many painters do precisely that. Painters are concerned with selling. The difference is that they are concerned with it after the fact. The designer, however, cannot afford to hold his concern until after the fact, for he serves a client who may never be free of that concern for twenty minutes.

The fine arts and the industrial arts may meet on the middle ground of the crafts, incorporating the purpose and freedoms of Art and the purpose and limitations of art. A carpet fabric by Jack Lenor Larsen is neither simply an esthetic experience nor simply a floor covering; it satisfies both functions simultaneously.

There is a depressing way to account for the current closeness of the industrial and fine arts. If they are more nearly equal than ever before, it is not because industrial designers are better but because painters and sculptors are worse. This is usually attributed to the art market, with artists accused of consciously shaping their styles to accommodate the prevailing fashion. (I can remember someone's remarking, "If the Museum of Modern Art announces an exhibition of figure painting, Tenth Street is full of figure painters where the day before yesterday there were only Abstract Expressionists.") They may, but that is not the problem. Bad painters don't drive good painters out of anything but the marketplace. It is not commercialism that brings painting closer to design, but the impoverishment of vision.

Also, the rules change. Life is still short and art is still long, but the former is getting longer all the time, while the latter gets so brief as to barely register. When the chips were down, commercial art could always be given second place because of its ephemeral quality; but in the face of the work of Chris Burden or Christo, the distinction got lost. A cereal box by law has to have a longer shelf life than the wrapping on a building.

The view is clouded further by the influence of art and design on each other. At one time Mondrian had so great an influence on American graphic design that any cigarette package with a couple of vertical lines and a crossbar was described as having a "Mondrian effect." The use of modern art in advertising is already an American tradition. In 1946 the Container Corporation of America began its "Great Ideas of Western Man" series, illustrated by such artists as Evergood, Léger, de Kooning, Moore, Lindner, Tamaya, Shahn, and Man Ray.

Container at least had a clear idea of its own motive. According to the design director, Egbert Jacobsen, "The series began as a forthright bid for attention to the company's developing business . . . and its awareness of the importance of good taste and top-notch quality in any public statement. . . .We do not think of ourselves as patrons of art. The fact is that patronage is the last thing in the world desired by artists of real caliber. As it is our business to produce containers, it is the artist's business to express ideas graphically and artfully."

Well, what happens when the work of a great poet and the work of a well known artist combine to push a box manufacturer? Does it sell boxes? Probably not, but then it doesn't have to. Container Corporation has to sell a great many boxes, but it doesn't have to sell them to a great many people. Theirs was one of the

The CBS eye designed by William Golden and the IBM logo designed by Paul Rand are calculated to work effectively on the printed page, on the television screen, and in the field.

earliest successful ventures into corporate advertising—advertising that doesn't sell anything more than a kind of fuzzy good will toward the corporation. This is a kind of Jewish mother advertising: look at all I'm doing for you, it says.

And what it's doing for you is pretty hard to disparage. Mobil brings you "Masterpiece Theatre" and J. Bronowski. Exxon brings you foreign films. IBM wants to share Galileo and George Bernard Shaw. Xerox in recent years has become a kind of International Endowment for the Arts and Humanities, having done as much for culture as Church and State and Ivy League.

A company is known by the company it keeps; at least it would like to be known by the company it would like to keep. Whatever Milton, Emerson, Whitman and St. John the Divine have in common, none of them appears to have expressed any interest in packaging. But a company that bothers to bring them to the world's attention must, it was assumed, be an extraordinary company. At a time when conformity was thought to be one of the most despicable drawbacks of American industry, here was Container Corporation urging everyone to march to a different drummer—or at least quoting someone who urged them to do that.

While at first this looked like a new use for art, an extension of an elitist activity into places where it would be accessible to everyone, it never really was that. The only corporations that could afford to use fine art were corporations that had to advertise without having any *thing* to advertise. (During World War II companies with no goods to sell used corporate advertising to keep their names alive.) There was nothing to lose and a world of image to gain. Thus art remained a luxury. Whenever fine art was associated with advertising at all, it was with the advertising of luxury goods: silver flatware, designer clothing, jewelry, imported cosmetics. But now there was a new kind of luxury advertising; because there was no selling burden placed upon them, the ads were a luxury for the corporations that ran them.

The idea of art as a luxury runs counter to any virile theory of industrial design. The most interesting thing about art as it relates to contemporary technology is its essentialness. Decorative art has always played a role in the development of useful objects, but that role seems to have been one of refinement; an impulse to go on and make a pitcher beautiful as well as serviceable. Graceless objects could be made more acceptable, more saleable, if ornament were added. It is easy enough now to laugh at elaborately decorated Victorian furnishings, but they betoken at least an awareness that art had a place in the creation of things.

Contemporary industrial design simply has a different view of where that place is. Above and beyond the maker's instinct for beauty, and the sales manager's instinct for something different and hence "better," probably only through art can we manage to live sanely with our machines. Design is a means of retaining our mastery over the machine and of forbidding its mastery over us. The industrial civilization that once relegated art to the museum would, paradoxically, be intolerable without a regular infusion of art. The oil refinery, the digital computer, the milling machine—each of these needs to be made understandable by art which comes into the factory not as an afterthought but as a necessity.

It is art that is the necessity, not the particular form that an object takes or is given. Many of the most beautiful forms produced by industry today are beautiful because their form is indicated, but not dictated, by what they are required to do. Musical instruments of the past have been admired for their appearance—an appearance defined in part by the operations to be performed on them. The way an instrument looks depends upon the sound it is expected to produce and the means of producing it—whether it is to be blown into, strummed against, struck, shaken, etc. Airplanes too are almost universally praised as examples of beauty that lies in purely functional design. Musicologists and aerodynamic engineers, however, assure us that some latitude in shape and appearance is possible for both instruments and planes. Obviously, an oboe could not look like a mushroom and neither could a plane. But the way they do look depends at least partly on an esthetic choice. Someone *decided* to make them look that way.

We begin to move away from Art with a capital A as soon as we take seriously the idea that form follows function. As I noted earlier, this is not a dictum to be slavishly adhered to. "Form follows function" is merely a reminder that it would be pretty dumb to do it the other way around. But even that moderate use of the principle suggests that we are not talking about Art. For in Art, form *is* function, or at least so nearly inseparable from it as to amount to the same thing. This is not true of industrial design. Even in its most Germanic sense, form tends to follow function only up to the point of designer's choice.

The motor for a power drill, for example, may be substantially the same as the motor used for other tools manufactured by the same company. So the motor's shape and composition are fixed. The designer is responsible for the housing. And the shape of the housing is not only functional, but expressive: it must express the interior mechanism and delineate the purpose and character of the drilling operation. As artist-in-industry the designer is supposed to see to it that both form and material meet these requirements. It goes without saying that the material cannot be prohibitively expensive, must be producible in the projected shape, must be resistant to the heat of the motor beneath it. But there is an additional requirement that the finished product "look like a power drill," and the decision as to what a power drill should look like is an esthetic decision. Function itself is not devoid of this esthetic consideration: part of any tool's function is to look like what it is. Otherwise we'd never be able to recognize one and would be deprived of certain pleasures and satisfactions in using it.

Those satisfactions and pleasures are also aspects of the design necessity. Most of us are separated today from the raw materials we depend upon. A lot of the most traditionally satisfying tasks are given over to specialists. The resultant tensions are obvious and one response to them is a reaction *against* technology, a conviction that machines are bad. This is more dangerous even than the Luddite position, which was based on a simple fear of job competition. Our own anti-technology bias is more complicated, and so are some of the fears it is based on. If industrial design becomes deeply effective, it can help make machine civilization more humane and technology less oppressive.

The incorporation of art into useful objects is most successful with products that are, as mentioned earlier, either "in" or "chic humble," extremely sophisticated and specialized or downright common and homely. A hen's egg is probably the best example of both: its elegance and beauty are undeniable; it is an everyday item, and it is engineered precisely to tolerances that let it pass through the hen's body without damaging it or her. Among manmade objects comprehensive examples are harder to come by. The clothespin, the safety pin, the paper clip, the two-by-four—these are all generic products that have achieved design excellence through evolution. More highly technological products are more likely to be beautiful by intention, although the intention is sometimes obscured. When the Museum of Modern Art in New York displayed a section of circuitry from IBM's RAMAC, many viewers questioned whether it was really art. The curator in charge defended its inclusion on the grounds that it not only represented genuine esthetic choices, but symbolized our changing sense of what constitutes an object.

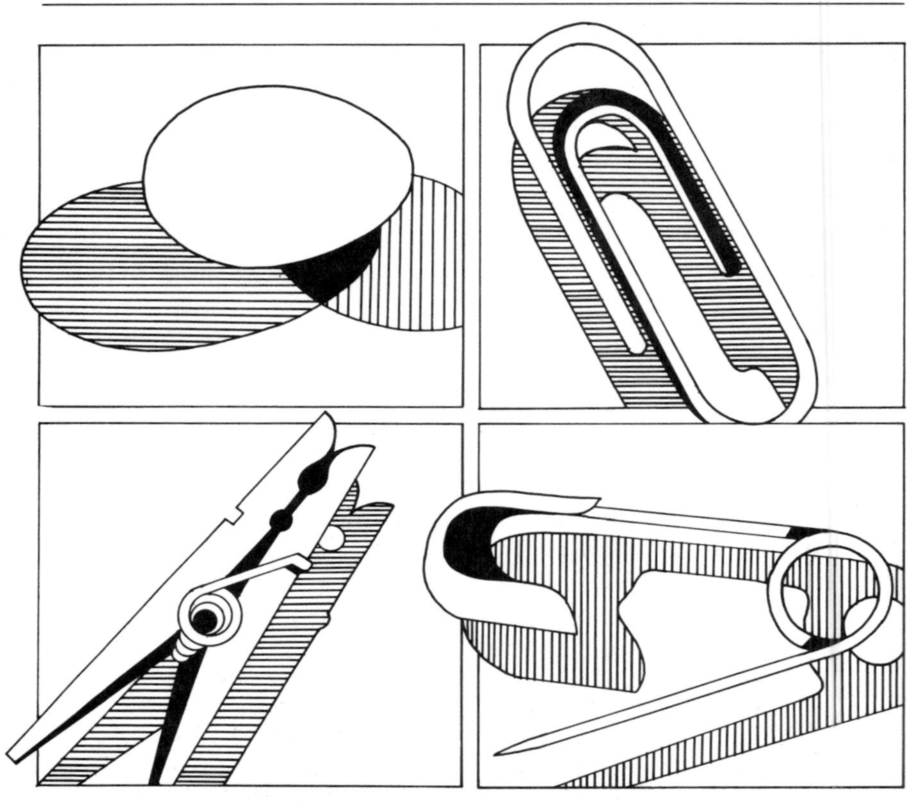

Superior design is often found in common objects.

There can't be any doubt, however, that some startling technological creations are only accidentally beautiful, as in a process introduced by General Electric for assembling tiny electronic parts. The components are set into a glass-base epoxy pegboard and flow-soldered. The resulting configurations are sometimes beautiful, but there doesn't seem to be anyone who especially wanted them that way, or who could have controlled them in any case.

The relationship between art and technology is not all one way. Many of the effects of modern painting and sculpture would be impossible without the materials that contemporary technology has provided. While this obviously includes improved pigments, it goes far beyond that when we consider electronic art, photographic art, and computer-generated art. Louis Sullivan wrote that "a science is sterile until raised to the level of art." In an industrial society it is not only sterile but dangerous. "The habit of art," Whitehead writes in *Science and the Modern World,* "is the habit of enjoying vivid values." The factory is beautiful by design. And if the habit of art is the habit of enjoying vivid values, the habit of creating industrial art is the habit of creating vivid values.

When an editor of *Product Engineering* magazine wanted to do a special feature on esthetics, he asked industrial designer Richard Latham to contribute an article. Latham declined, but wrote a letter instead, about why he didn't have much to say on the subject. Here is one of the things he didn't have to say.

> Before the advent of the Industrial Revolution, the things that were large enough and solid enough to stay around and dominate their environment were mainly architectural. Cathedrals, pyramids, medieval towns, Roman aqueducts, etc. These large masses of stone and mortar remained to fill up the landscape and in a very real way affect man's surroundings esthetically on a permanent basis.
>
> Since the Industrial Revolution things have changed considerably. The machine has grown up fast. Not just in size alone, but in numbers and to an extent that it affects our esthetic sense twenty-four hours a day, and has begun to dominate its environment.

The dominance of machines may seem less formidable when we consider that technology rests upon disciplines that are not unlike art in the way they operate. In a paper given at the University of California in 1946, John Von Neumann observed that the mathematician's "criteria of selection . . . are mainly esthetical," and that they have this in common with the criteria of theoretical physics. Granting that mathematical ideas may originate empirically, Von Neumann said that "once they are so conceived, the subject begins to live a peculiar life of its own and it is better compared to a creative one, governed by almost entirely esthetical motivations, than to anything else."

Mathematics is probably closer to Art than commercial art is, for like Art, mathematics has no purpose outside itself (although it can be used for many purposes). Commercial art always has such a purpose. No matter how closely commercial art may resemble, or even surpass, fine Art, it still has an extra-art job to do. It has to sell something. At least that is the tradition. But what happens when Art takes its cue not from life but from art?

This was the case with pop art. It may at first have looked like a replay of Dada, but its attitude was vastly different. Dada was satirical but also admiring. And with good reason. If you're copying comic strip artists and package designers, the least you can do is acknowledge their talents. I remember someone's looking at the early work of Roy Lichtenstein and saying, "He's good, he's really good. But you know something? The guy who draws Steve Roper is also really good."

Once pop artists began putting the images of commercial art into galleries and studios, it wasn't long before commercial artists found ways of getting there directly. In addition to the galleries and museum shows mentioned earlier, a wide variety of commercial productions—including orange crate labels, transparencies from *The Yellow Submarine* and of course posters—have been sold, and at good prices, through magazine advertisements. And while Ethel and Robert Scull were buying Art derived from comics, comic book illustrators were selling their wares to fans through direct mail and at comic book conventions and nostalgia fairs.

Yet commercial art has remained an embarrassment, partly because the lines between upper and lower case art become so increasingly fuzzy where they are even discernible. The excellent magazine *Communication Arts* has from its inception been known as CA, but that abbreviation originally stood for *Commercial Art.*

Writing in CA *(Communication Arts)*, Milton Zolotow (commercial artist) describes his days studying fine art at the Art Students League: ". . . there we learned that usefulness and 'art' did not mix. We learned this on canvas because canvas was an enduring material and we were addressing ourselves to posterity." That was what happened on the top floors of the League. "On the lower floors of the Art Students League," Zolotow recalls, "we studied commercial art and there we learned that usefulness and art did mix. We learned this on illustration board because we were addressing ourselves only to the photoengraver."

After the art market boom of the sixties, however, Zolotow observes: "Soon it became hard to tell one breed of artist from the other. They both wore denim and leather at the Italian movies and began to steal each other's tools. The commercial artist-painter found photomechanical tools very handy for the production of commodity paintings in large quantity. Why paint Elizabeth Taylor when you can silkscreen her image on canvas with a line resolution? Why not project a photo on canvas if photorealism is your goal? Meanwhile the artist/commercial-artist shares the same acrylic paints and glaze mediums, makes woodcuts, collages, sculptures, engravings, and just plain drawings for the engraver's camera."

An important traditional difference between the commercial artist and the other kind was that the former was almost always anonymous, with exceptions like Norman Rockwell and James Montgomery Flagg. Knowing the name of a commercial artist was as improbable as knowing the name of the copywriter for a dog food advertisement. That's no longer true. Graphic artists like Milton Glaser, Ivan Chermayeff, R. O. Blechman and Paul Davis are deservedly well known. Chermayeff's massive, three-dimensional 9 on New York's 57th Street is, if not a landmark, at least a tourist attraction. Art students from Kansas City make pilgrimages there. It isn't just a sign. It isn't exactly sculpture either, but it must be at least as close to it as Robert Indiana's *LOVE*, cheap copies of which can be found in souvenir stands all over the country, a case of life's not only imitating art but ripping it off at great profit and with considerable poetic justice.

Chermayeff, a partner in the graphic design firm of Chermayeff & Geismar, says, "If cab drivers ask me what I do, I say I'm a commercial artist. It's just simpler. I think that Roy Lichtenstein is a commercial artist too. If you can knock off a certain number of paintings because of making money, you're a commercial artist. If someone pays for it *before* you do it, you're doing it for a reason. It's somebody else's reason and responds to someone else's problem."

Number 9 designed by Ivan Chermayeff for 9 West 57th Street.

Milton Glaser, a founder of Push Pin Studios, sees it pretty much the same way.

> Fine art deals with internally imposed problems. But if there's no external problem, there is no design. Compromise is accommodation to the external problem. Art and design come closest when the nature of the message is ambiguous. It's not really a difference between high forms

> and low forms. It's really a difference between art generated by artists who "own" forms and art generated by artists who use those forms. But if the external problem is to get into a gallery, then you are a commercial artist whether you are a painter or not.

Perhaps because commercial art is always done for a fairly specific purpose, it is unlikely to be effective when harnessed to another purpose or to none. Misusing it is comparable to misusing any other tool, like trying to hammer with a screwdriver. It doesn't debase the screwdriver, although it may break it, but it doesn't do the job very well. In the absence of a specific purpose, fine art is far more vulnerable. Ironically it can be much more useful, but not more powerful in this world. Because it has no clear-cut job description, it can be pressed into service easily—particularly if, like Abstract Expressionism, it takes a form that denotes nothing and that connotes not much more. But the power of commercial art is like the power of money, to which it is related. Because it is what people see, commercial art helps determine *how* they see. The "visual illiteracy" of which designers complain is very largely a result of the "visual pollution" they helped to create. Like most pollution, it is a by-product of an uncontrolled process.

For the power of commercial art is largely uncontrolled. Because their objectives are concrete and practical, designers may appear to have more choice than they really have. Painters may rationalize their work, and too often do, but designers *have to,* and are therefore prone to overlook what artists never would: the significance of talent. The public is also willing to overlook it. A hideous painting is attributed to a lack of painterly talent rather than to a sinister plot against us. But a hideous package design is taken as a personal attack, as if the designer had it in his power to do better. Our worst commercial art is not the result of evil intention but of inadequate talent.

Like serious painters, designers develop styles and go through periods. Once I was driving across the Rocky Mountains with Deborah Sussman, a West Coast graphic designer. As we drove by an outrageously vivid collection of wild flowers, Deborah said, "Alexander Girard did that." A mile or so later, passing a stern rock formation, I said, "Gropius did that." That's how the game began and we played it for about half an hour with each of us challenging the other to identify the authorship of sunsets, trees, boulders, and streams. That was fun, but what does it prove? Nat Fleischer, of *Ring* magazine, could do the same thing with boxing styles, and do it better.

Still, the existence of stylistic distinctions means that, for all the anonymity of commercial markets, the design process yields to a personal stamp. This, to be sure, could be just the triumph of the designer's ego. But at its best it can mean the assertion of humanness where we forget to expect it. That assertion must be found in the techniques of an industrial society, rather than occasionally exhibited as a cultural alternative to it. And capital A Art needs lower case a art for, as Jacob Bronowski warned, "If you neglect the seed ground of a lively industrial art, then all art withers."

The Design of Possibilities

"Nobody smokes in church." Richard Farson

If nothing is so powerful as an idea whose time has come, nothing is so enervating as an idea that's been sitting around for years like money not earning any interest. That is precisely the situation of "situation design," the concept of moving from the design of *things* to the design of the circumstances in which things are used.

As far as I know, the notion that situations might be designed by professional designers was first mentioned in the early sixties by Edgar Kaufmann, Jr. The phrase "design of situations" has surfaced periodically ever since, unsupported by much explanation of what it might reasonably mean. Examples that *have* been given are invariably anticlimactic: to describe as situation design the arrangement of a room in which an important meeting is going to take place is not inaccurate, but it hardly suggests a radical shift in design significance.

Yet situation design was being heralded as significant. Designers were at last on to something new and serious, if there were only some way of pinning down what it was. If designers couldn't tell us, maybe fellow travelers could.

Richard Farson is a psychologist with an interest in design, particularly in the paradox that the more design we have, the more problems we face. (*He* thinks this is a paradox; some of us find it perfectly natural.) In a 1966 paper dealing with social implications of the human potential movement, Farson argued that professional designers were needed in the design of human situations. "What will be the situational designs that will help to make us really healthy?" he asked rhetorically. No one knew. In the meantime, though, Farson observed that environmental design was already controlling situations in ways we had come to take for granted. "Nobody smokes in church," he reported, with an enthusiasm ordinarily reserved for the discovery of new planets.

That relevation might seem dazzlingly irrelevant to anyone but a priest or a fireman, but Farson had a point: the church designed the behavior of the churchgoer. The church also designs the behavior of the minister, and astute clerics have always understood the importance of church design. Traditionally they have understood it much better than playwrights have understood the importance of theatre design. This has changed in our time, and more flexibility in theatre design on the one hand and the reemergence of street theatre on the other have combined to create object lessons, not always exemplary ones, in situation design.

The Ford Foundation once sponsored a program to bring designers together with playwrights and other theatre people to talk about concepts for the "ideal theatre." One of the designers, architect Pietro Belluschi, concluded that "theatre houses impose severe physical and artistic limitations on all types of performing arts by the inadequacy of their design. . . ."

The Ford project consisted of eight design concepts for the "ideal theatre." The difficulty with such a project is inherent in the statement of it, for there is no ideal theatre. More important, there is no prospect of designing one, except "conceptually," for such a project is free of the constraints that make design possible. Some truly top-flight architects and theatre designers worked on the Ford venture, and presented their concepts. The most interesting design commentary, however, came from playwright Arthur Miller, who said "I have no doubt that plays are not being written just because of the limitations of New York's theatres. . . .You can't write for those 'shoeboxes' with the same ideas, with the same emotional scope, as you would for a [more adaptable] theatre. . . . You can't hope to make one theatre which is absolutely perfect for all kinds of plays."

Similarly you can't design one type of environment for all situations, which is why situation design has to be constantly called into play. When churches prohibited smoking, restaurants supported it by selling cigarettes and providing ashtrays. Today restaurants and other public places are pressured to discourage, even to prohibit, smoking, while churches are less likely to find it sinful.

Environmental design can direct human behavior, but it isn't easy. At the United States Conference on Human Settlements, held in the summer of 1976, the Greek and Iranian delegations proposed resolutions on design, urging that housing be sensitive to "human scale" and to regional lifestyles and building traditions. The Greeks went further, calling for "the creation of chances for human encounters and the elimination of urban concepts promoting human isolation."

It seems unlikely that such changes can be effected by, or even affected by, resolutions and declarations and conferences. Sure, some urban concepts promote human isolation, but not because an international forum has failed to warn planners to get with it.

The Canadian architect Irving Grossman tells of designing a housing project for the elderly. Compassionate and conscientious, Grossman was determined to give every resident a view of the green park enclosed by the project. And he managed to do that with most apartments. At last, though, he admitted that, to stay within the budget, some apartments—as few as possible—were going to have to look directly out on the superhighway that ran by the complex. Defeated, Grossman shared his despair with the contractor, who said, "Don't worry about it. Those apartments will be the first ones to go."

When they began to rent the space, the contractor turned out to be right, for he knew what Grossman hadn't known: the elderly tenants felt isolated enough already; they didn't want a peaceful green nothing to look out on (with other elderly people sitting in it). They wanted to be, at least visually, where the action was.

Industrial designers, who have always welcomed support from any quarter, were pleased to be perceived as moving from the design of things to the design of situations, whatever it meant. By the sixties, designers generally had become as bored with products as educators had with teaching (a survey of products from the period will show how bored they were) and were beginning to describe themselves as planners, consulting generalists, conceptualizers—always in language that kept its distance from the world of objects. "The product is dead," a Chicago designer told me as we sat in his Alfa Romeo. He intended to devote the rest of his life to pure process. "I have styled my last hoo-ha," said another. Others, acknowledging that they could not so easily break free from the bondage of goods, began speaking of "total design," implying that the object was at least kept in its place—a place usually subordinate to marketing.

Situation design, then, was an appealing concept rhetorically. It was dignified, sounded significant, and required no special professional training, but only the amorphous, comprehensive array of talents that all designers were presumed to have. Yet if you asked people, "Designed any good situations lately?" you waited a long time for an answer.

In point of fact designers do not often design situations professionally. Then who *does*? Novelists do. Comedians do. Playwrights and directors, marriage brokers, football and basketball coaches, group therapists, lawyers, politicians. . . .Situations are being designed all the time, but since they are not designed by professional *designers,* no design credit attaches to them. So there really is nothing new about the design of situations as such; what is new is the idea of approaching them with professional design skills.

Some situations are important and complex enough to demand whatever insight and expertness we can find. The response to an unexplained plane crash used to be an investigation of what was wrong with the plane; thus we discovered metal fatigue. Aircraft design defects and excessive material stress are still possibilities, but today we mean something quite different by "what went wrong?" We are beginning to look at the design of the situation—communications practices, certification procedures for aircraft, traffic control procedures, physical examination of both aircraft and pilot, landing strip configurations, lighting systems, etc.

Actually we refer continually to situation design in our everyday lives, but we think of it as just a figure of speech. Following President Sadat's assassination, the Israeli ambassador said, of Sadat and Prime Minister Begin, ". . . these two major architects of the peace process intended to create a situation that would outlast them." This language startles nobody; nevertheless, we do not really think of this as design. I turn on the television set Sunday morning and find myself watching a religious drama broadcast by the Paulist Fathers: a probation officer is trying to console a depressed prostitute. "You build a life like a building," he tells her, "day by day." The metaphor is sound enough, but it is more than a metaphor. That *is* how you build your life; and lives, like buildings, are designed or they won't stand up.

In 1954 Peter Drucker introduced a business strategy he called "management by objective," which has since become highly influential. Now one Richard L. Wessler, Ph.D., offers an approach to life called "*Self*-Management by *Personal* Objective" (italics mine). He introduces it for a fee of $20 in a seminar called "Designing Your Future."

When, with his company facing bankruptcy, an official of Braniff Airlines said in a television interview "we're going to have to redesign the jobs and the work process," he was not speaking figuratively: redesign was needed. He was, however, speaking too late.

The actor and director Robert Redford has recently established the Institute for Resource Management, through which he intends to bring together environmentalists and corporate executives in energy-related businesses. To be effective, the Institute will have to do more than simply get Barry Commoner and the head of a utility company in the same room. That kind of thing already happens and at best makes for good talk shows. The challenge is to go beyond dialogue to the design of situations that lead to action. Redford has also set up on his ranch in Utah the interdisciplinary Sundance Film Institute at which actors, writers and directors can develop sensitivity to each other's crafts. And he has done this very largely as a design project, with the help of professional designers.

The traditional design of situations lets us examine some aspects of design process in the absence of design agonies. Take comedy for example. When I was a child we had a device in our living room that brought in the world's best entertainers for nothing but the cost of exposure to a few commercials. It was called radio. I used to listen to Fred Allen, Jack Benny, Doc Rockwell, Fibber McGee and Molly without realizing what made them so unfailingly interesting week after week, until one day my father said, "Did you ever notice that the best comedians never tell jokes? They create funny situations."

My father didn't know he had uncovered a design principle, but he had: telling jokes is the styling of comedy—relatively easy to master, accessible to the less talented, likely to follow trends and fads. Polish jokes, moron jokes, elephant jokes, lightbulb jokes—these are the equivalent of tail fins and safari suits and corporate logos full of arrows. The one-liner is always sharply limited in range, no matter how marvelous the execution. There is a sense in which Bob Hope is the Raymond Loewy of comedy. Bob Hope is very good at being very funny, but he does not lead to any understanding; there is nothing for us to *do* about Bob Hope—as with styling, the consumer is passive and uninvolved. But Chaplin shows us something of what it means to be a person.

In the hands of a designing comedian, if a joke is used it is generally recast as a situation. Comics don't say, "I am reminded of a story . . ." or "Did you hear the one about? . . ." Rotarians do. Comics say, "A funny thing happened to me on the way. . . ."

Nor do true comics say, "But seriously folks." That's another clue to styling: the shift in character, the abrupt break when the joking stops and the pitch for the March of Dimes begins. True comedy never has to step aside for the serious. It *is* serious. That's why it's so funny. Design can be funny too—deliberately, as in the Olivetti typewriter case that becomes a wastebasket, or unintentionally, as in the Aurora "Transition" caskets with snap-in decorative panels to give the user a choice of pastels or as in the streamlining of a pencil sharpener or a tea kettle, although neither benefits from lowered wind resistance. The Bauhaus didn't take humor into account and the Post Modernists threaten to take nothing else into account. It is fallacious to think that everything has to be either funny *or* serious.

Every basic change in structure requires a corresponding change in behavior. That makes situation design necessary. What makes situation design possible is that the structural problems of any given client are not as special as the client thinks they are.

This is illustrated dramatically in every design project that has to do with communications. If you are designing an exhibition or a film, or any piece of corporate literature more intellectually complex than a sales brochure, people in the company keep saying, "I don't envy you." What they mean is that the particular problems of their own industry or company are so difficult and special that no one outside the organization could possibly understand them. If you do understand them, or appear to, they are amazed. Of course *you* know that your grasp of their problems is based not on your astuteness but on your experience with other problems very much like theirs. If you tell them this, they are either incredulous or hurt.

It is the commonality of human problems that makes feasible the practice of fortune telling, medicine, and every kind of outside consulting, including design. An experience of George Bernard Shaw's illustrates this.

> The first time I had my hands examined by a palmist he amazed me by telling me the history of my life, or as much

> of it as he had time for. Apparently he knew about things I had never told to anyone. A few days later I mentioned in conversation with a friend . . . that I had been dabbling in palmistry. He immediately put out his hand and challenged me to tell him anything in his life that I didn't know from my acquaintance with him. I told him about himself exactly what the palmist had told me about myself.
>
> He too was amazed, just as I had been. We had believed our experiences to be unique, whereas they were 99.9% the same; and of the .1% the palmist had said nothing.

To an anatomist, Shaw concluded, monkeys are all skeletally alike, with the exception of a bone or two. This tends to be true not only of monkeys and of people, but of organizations, both commercial and nonprofit. (Of course, much of the excitement in design comes from dealing with the unique bone.)

No matter how much they are alike, however, situations are difficult to design. A situation is thought to be static, something you're caught in—a pickle, a jam, a state of affairs. ("That's the situation and the box it came in," Constance Bennett tells Roland Young in *Topper.*) But a situation is static only for descriptive purposes, just as posture seems to describe a fixed position but is rarely encountered in the absence of movement. (An exception is the military posture of standing-at-attention, which is both barbaric and physically harmful.) A football coach may describe a situation symbolically on a blackboard, but usually he is expressing motion as well as position. Situations are dynamic, like the design process itself, in which fixed focus can be crippling. I knew a very successful and very good industrial designer who became obsessed with the problem of writer's cramp, which he was convinced was caused by the improper design of pens and pencils. A serious student of human factors, he spent years studying the muscles of the hand to determine the basis for a new kind of writing device that would permit long periods of writing without strain. Although hardly of epic scale, the problem was serious enough, and there was no disputing the designer's diligence and responsibility. If ever there was a right way to go about designing a product, this surely was it.

But when the Japanese introduced the felt-tip pen, soon followed by American imitations superior to the original, the problem shifted radically. There is now so little friction in writing that writer's cramp has been virtually eliminated, even though our pens and pencils may be as hard to grip as ever.

Perhaps the most valid model for the designer of situations is the scientist, because of the open-endedness of scientific experiments. The experiment itself is designed, but the working rules require that the designer not manipulate the process. The scientist sets up a situation on the basis of reasonable prediction, but in fact he does not know what will happen. Neither does the designer.

The assumption that designers *control* situations leads to self delusion and also to the delusion of clients. Manhattan's office building plazas—populated by bums, prostitutes, and ambulatory psychotics—are built from architects' models made credible with the aid of nicely dressed figures sitting still, admiring the fountain and generally making the scale of the building look tolerable. The trouble, as William Hyde Whyte's film studies of behavior in public places show, is that people don't behave like the cardboard people in architects' models, because what the cardboard people don't do is behave.

Do we need better instruments for modeling behavior? Perhaps. But the problems are social, not technical, and we need social designs to address them effectively. The most elegant design solution of the fifties was not the molded plywood chair or the Olivetti Lettera 22 or the chapel at Ronchamp. It was the sit-in. Achieved with a stunning economy of means, and a complete understanding of the function intended and the resources available, it is a form beautifully suited to its urgent task. The form did not pop into existence with someone's spontaneous refusal to sit in the back of the bus. It was the conscious creation of strategists like Bayard Rustin and, years later, Martin Luther King, Jr., who adapted Gandhian protest techniques to Western problems.

It was also ideally suited to the college campus. A student body is, among other things, a collection of young people artificially segregated and confined. In the distant past the animal spirit not taken care of by sex and spectator sports erupted in the form of riots, panty raids, goldfish eating, and pranks. For generations faculty and staff nervously endured student pranks, hoping it wouldn't be *their* car that was painted with red stripes and moved to a no-parking zone, and devoutly wishing all that energy could be put to constructive use.

Suddenly it was. The demonstrations and confrontations that came to a head in the late sixties were student pranks raised knowingly to the level of social responsibility. However cruel, the practical joke has always been a way of designing situations. And what defeated political adversaries was the same undergraduate design ingenuity that had defeated deans and dormitory house mothers and building custodians.

So the theatre manager described on page 7 was perfectly correct in accusing the demonstrators of carrying out a prank. In planning the sit-in, the students were advised and aided by experienced, dedicated professionals who were instantly dismissed as "outside agitators." (Not always unfairly; that is what designers have to be sometimes.) They had raised such basic design questions as: What do you want to accomplish? What are the materials you have to work with? Who are the people involved? What do they need and want? If this is a problem, what might a solution look like? What forms might it take? What form can you give it?

The forms they designed appropriately addressed the needs of all concerned. Blacks needed equal access to the theatre; students needed to experience their rights and powers; the town needed to confront the issue; the theatre manager needed an out, an excuse for complying with the law, a dramatic demonstration that his hands were tied. For, as it happened, he *was* personally indifferent to who sat where.

Physical protest had been tried before at that theatre and at restaurants throughout the area, but the simple refusal to move had accomplished nothing more than temporary disruption. There had to be a *strategic* refusal to move, with the odds arranged in favor of the designers. (As with most design, there was some risk: people could have been injured or arrested. In other civil disobedience designs, arrest is itself a strategic device. "Fill the jails," admonished that master architect Gandhi.) Buying the tickets in advance assured surprise and control. The carefully planned takeover of the theatre, with each student knowing where to go and with whom, consolidated the advantage. And the press release and letters, prepared weeks in advance, locked the manager into a course of action. By removing the affair from the sphere of negotiations, the designers had made him an offer he actually couldn't refuse, for there was no place to lodge a refusal.

That happened in the 1940s! And the people who had made it happen, in conjunction with similar experiments across the country, developed techniques for desegregating restaurants, stores, schools and other institutions, and perfected the techniques in workshops and clinics. If that theatre sit-in was a clever prototype, the sit-in of the fifties and sixties was a highly refined instrument with most of the bugs worked out. And by the seventies the design had been copied, marketed greedily and thoughtlessly, used as if it were a cure-all instead of a specific, and rendered ineffective much of the time. Anti-war demonstrators designed the situation of the 1968 Democratic convention, with a little unwitting help from their friends the police and from city officials. Because the Chicago police had no experience in design, they capitulated; although it didn't look that way on television. Yet how-it-would-look-on-television had been incorporated into the design proposal from the very beginning.

Television *demands* the design of situations. More than 97 percent of the American population live in households with television sets, and for the majority of those people this still new, unprecedentedly powerful and blatantly imperfect medium is the chief source of information about what's going on in the world. Consider the design elements in the CBS evening news.

The program format is designed, with time slots carefully allocated to major news, minor news, commercials, windups. And it is, even in public television, a competitive format, with the resources of each network dedicated to getting people to turn to their channel rather than another. But within that format, each day's program is designed by reporters and writers and by the director in the control room.

The news is a result of an elaborately designed mechanism for capturing and editing events on film and videotape, an exercise in the logistics of camera crews, and continual redesign of equipment for improved performance and portability. The studio must be designed to accommodate a large support staff to supply the people on camera with what they need.

As for the on-camera correspondents, they inhabit what really is a set for a nontheatrical play, raising scores of design questions. Does an anchorperson need more than a table and a chair? What is the appropriate tone? Should the furniture be ultramodern to connote an up-to-the-minute operation or traditional to connote reliability? Should the anchorperson shuffle papers, suggesting that he has and records his own ideas, or should he be backed by an armada of monitors, flashing lights, ticker tapes and whirring computers? *Star Trek* or the city room of the *Washington Post*? If the desk and chair are part of a set, what does that make the anchorperson? Not exactly an actor but plainly not a backroom journalist either. He is on camera, wears makeup and, because TV is a visual medium after all, is chosen at least partly for the way he looks and comes across.

The set and the cast and the news footage are not the only elements that appear on the screen. There are the graphic and typographic support elements that some designers regard as "wallpaper." When Dan Rather is talking about the national budget deficit, should a dollar sign be displayed on the screen? Or the word "Deficit," accompanied by the latest figures displayed so viewers will have a better chance of grasping them? Or a picture of the Treasury Building? Or the President with his budget director? Or the whole budget team crunching numbers? (What does number crunching look like?)

Behind all this of course is the design of the news-gathering situation. An interview is usually a meeting designed by a reporter. A press conference is usually an event designed by an official or his staff. The press conference is a good format for delivering what the deliverer wants said, but not a very good device for finding things out. Good reporters try to accept the constraints and to design questions that will elicit useful answers anyway.

And this is the tip of the iceberg the anchorperson rides.

When Walter Cronkite was succeeded by Dan Rather, CBS designers wanted to redesign the physical surround, but didn't get it done. For awhile they were introducing the show with a computer-generated image of the world. This, however, took 13 seconds of precious time and did not seem to enrich the program. Design vice president Lou Dorfsman and his associates replaced the globe with what Dorfsman calls televised headlines—a series of brief cuts of major news stories. The problem was how to make these special without making them gimmicky. By the date scheduled for the change they had no solution ready, so Rather simply read the headlines. "When I looked at it that night," Dorfsman said, "I realized that was what he *should* do. The most effective design in this case was no design."

Museums, whatever their content, are logical design arenas. Their renewed vitality reflects a spreading curatorial perception that a museum is a designed situation more than it is a warehouse open to the public. This in turn has made it possible for a great many people, including children, to perceive museum-going as something to do, rather than something that is done to you. At the San Franciso science museum called the Exploratorium, the experience is enhanced by the incredibly learned and inventive young aides who uncannily appear just when a visitor wants help and seem to vanish as the need diminishes.

When the Cooper-Hewitt Museum opened in 1976 I took my children to the opening exhibition, which included a series of doors. The doors themselves were not all that interesting; what was interesting was that you could open and walk through them. The fact that we all have such doors in our homes did not seem to diminish the delight of finding them in a museum. Several weeks later I suggested to my nine-year-old daughter, Leah, that we go to a show I thought she would enjoy at the Whitney Museum.

"Are there any doors you can go through?" she asked.

"I don't think so."

"Are there parts of it that you can use yourself?"

"I doubt it."

"Then it's not as good as the last one we went to," she said resignedly.

Martin Friedman, director of the Walker Art Center in Minneapolis, says "I agree that our museum is a perfect example of situation design." By way of illustration, Friedman points to the Walker's guards who, dressed in gray pants and casual blue jackets, seem to be there to enjoy the pictures. "They don't look like centurions," he points out proudly.

"We think not only of designing an exhibit, but of how it will work, how it will be perceived and used," Friedman explains. "We try to give people the necessary background, the context, the relationships, so we have an information room—a small theatre for showing movies and slide shows relative to, but not part of, the exhibits. The game lies somewhere between the need to interpret and the commitment to present, and we always try to interpret in a way that is connected to experience."

College professors plainly ought to be designers of situations, but they rarely are. One would expect educational administrators to stand in relation to teachers as corporate executives do to professional designers—as clients. But there is no agreement about what the educational product *is*. Administrators talk as though *students* were the educational product. ("We turn out a well-rounded liberal arts major here"). Students are not the product. The only educational product schools can be reasonably charged with designing is the educational environment—not just the classrooms and dormitories and recreation centers that college presidents dedicate their energies to acquiring, but the situations in which students interact with each other and with faculty members. Increased costs and decreased federal support make the effective design of these situations a matter of some urgency.

Before World War II, and in some places for a full decade afterwards, liberal arts colleges *offered* a course of study but *provided* something quite different. For four years you could ripen on a tree-lined campus like old sherry in a cask. The History 101 textbook at Oberlin might have been identical to the one used at Wayne State University in Detroit, but the situations were vastly different. The history text was (as any textbook is) among the least important components of an education, and could as easily have been read on a Greyhound bus as in a dormitory. The *four-year environment* was the real text. Colleges were in the situation design business, whether they wanted to be or not, and faced circumstances that were familiar to industrial corporations: the inevitability of change coupled with the longing to stay just the way they were.

Uneasily, they turned to technology. There was good precedent. Hadn't medicine, faced with the ticklish problem of diagnosis, been saved by antibiotics? Educators desperately wished to believe that the new educational technology, like penicillin, would be indiscriminately applicable to any ill. Closed-circuit TV, film, computer-assisted instruction, programmed teaching materials were welcomed as the antibiotics of education, empowering us to make things better without having to figure out what was wrong. Few cures were effected. New technology never cures anything, although it can sometimes make cures possible. That is where Marshall McLuhan comes in, and why he used to come in so often. However reckless some of his provocative pronouncements may have been, McLuhan made a valuable nagging contribution: he kept reminding us that the new media are not the old; that television is not a book or a lecture.

Technological capability tempts us; but the fact that a biology lecture can be put on videotape doesn't mean that it is worth doing. The fact that a computer could store an Oxford tutorial (and for that matter wear tweeds and smoke a pipe, if anybody wanted it to) doesn't mean there is necessarily any point in using the computer that way. But if there is no point, there is at least some explanation. New technology is at first guided by old technology, and educational technology has been no exception. If it is understandable that the first automobiles were modeled after buggies, it is even more understandable that the new educational devices should be modeled after books and blackboards, and programmed to support lectures.

Lectures were crucial when books were rare and costly. The man who owned the book gave the lecture just as the kid who owns the ball is captain of the team. By the nineteenth century the book-lecture sequence was beginning to reverse itself, and books were becoming devices for reaching people who could not attend lectures. The contemporary undergraduate lecture is a device for reaching students who do own books but cannot be expected to—and in many cases are unable to—read them.

In that sense, students need lectures; but that is not why we have them. We have them because the faculty need them. Apart from ego considerations, a lecture provides the narrative thread for a course of study. Lecturing is "what teachers at the college level do."

They do not necessarily do it well, and don't have to. The ability to teach is almost never related to anyone's entry into, or advancement in, the teaching profession. The old joke is that those who can, do; those who can't, teach. The *real* joke is that those who can't teach, teach.

But an indifferent lecturer could in many cases be a superb member of an instructional design team. It may not be the highest calling in the world, but the role of instructional designer is superior to those of disciplinarian, scorekeeper, monitor, juror, scapegoat, knowledge broker and other roles that teachers conventionally play. It might even pay better.

The necessary collaboration will be difficult to achieve. Although many exhibitions, films, brochures, and other designed projects are broadly "educational," they are not required to assume the instructional burden carried by a university course. And although designers know how to create adult show-and-tell materials, most of what they show and tell is superficial—primarily because superficial messages are what their clients wish to deliver. Could materials of depth and substance be communicated by the same messengers through the same technical media? Certainly, but not in the same modes of discourse. Advertising jingles may sell toothpaste, and posters can announce anything from a new movie to the second coming, but that does not mean that jingles can teach history lessons or that posters can describe nuclear theory. Where is the agency to handle the instructional account?

There is none; it isn't a job for an ad agency, not even an in-house agency. That is why establishing campus "audio-visual centers" will not do the job.

Centers are *in* these days. My neighborhood has a nutritional center, a hair styling center, a yoga center, a center for the performing arts, a family dental center, a vision center and two copy centers. In the suburbs these would all be located in a shopping center. With so much centrality around, it's hard to know where the edges are.

Campuses do need facilities for producing slides, processing film, and arranging video hookups, but "centering" audio-visual activity is to render it peripheral. There is no creating exceptional educational materials without the intense collaboration of academic experts. To hand subject matter over to media experts is to invite packaging when redesign is what is called for.

An experimental way to start an instructional design program might be the redesign of, say, a one-semester course in a difficult subject. The object would be to devise a course demonstrably superior in effectiveness, efficiency and pleasure to the course it replaces. To achieve this the same design team would have to be responsible for every aspect of the course from content organization to laboratory equipment.

Professors don't have time for that sort of undertaking; they are too busy not teaching. We can hardly blame them for resisting new materials and methods, especially when the materials have been grossly oversold and the methods untried, and college teachers are not rewarded primarily for instruction in any case. Dean Henry Rosovsky of Harvard reminds us that "there is no Nobel Prize in undergraduate education."

When a behavioral psychologist demonstrated computer-assisted instruction to a group of college professors, one of the audience objected, "The trouble is, there is no crotchety old sonovabitch in

In his design of a Montreal drugstore Francois Dallagret included this surrealistically chic fast-food restaurant, a womblike environment intended to "breathe." At night it became a discotheque.

that machine." He was right, of course; but the trouble is, there usually isn't a crotchery old sonavabitch in the classroom either. Teaching machines *are* unimaginative, but so were most of your teachers, remember? The situation permitted it. It will not permit it any longer, and has to be redesigned.

Interior designers are in a position to design situations, and sometimes do, although too often they do it stereotypically. Think, for example, of all the restaurants that are scientifically unlit, because candlelight and other forms of darkness create "intimacy," although they make dining difficult and the only intimate behavior they encourage is a kind that's hard to manage during a meal anyway. Their ultimate effect is less the promotion of intimacy than the exacerbation of separateness. Dark rooms and soft music are all a college boy needs to know about the parameters of seduction, but design research ought to have gone further than that by now.

A small midwestern town I know has a sun-drenched, outra-

Restaurant interiors offer latitude for situation design. This fast-food-counter restaurant pioneered specially designed devices and processes for speedy hands-off service.

geously public storefront bakery with a serve-yourself coffee bar, a few tables for snacks, but no booths. Far from romantic, the ambience is calculated to militate against even the subtlest flirtation. But one winter day I saw a couple there so deeply lost in each other that they were oblivious to the decor, the street traffic, the people around them, and even the pigs-in-blankets they were eating. Nothing in the way of dark woods, soft lights, or mood music could have added anything but what communication theorists call "noise."

It is more promising to look at interior design as a means of improving work. Not just productivity, in the efficiency expert sense, but pleasurable achievement. The so-called "office landscape," or open office, was introduced with great fanfare in the fifties, although no introduction was necessary to anyone who had worked in a newspaper office or seen a movie about one. Almost any old movie newsroom shows the advantages (low cost and ease of communication) and the disadvantages (noise, loss of privacy)

Considering that more than half of the workers in the United States spend one-third of their days in offices, relatively little attention has gone into making them effective workplaces.

of the arrangement.

Most offices are equipped with tools that do not look like tools, and so we lose sight of the fact that tools are what they are. Desks and office chairs, for example, have not so much evolved as happened. And the capriciousness of their happening has been compounded by the tendency of interior designers to make executive offices pleasant by treating them as parlors. Although this is a mistake, it is rarely a tragic mistake: most living room furniture is about as well suited to office work as most office furniture is.

It is true that performance criterial have been inescapable for certain kinds of office equipment, and such equipment has been improved and refined. Ironically it consists chiefly of tools for performing relatively unskilled tasks. Typewriters have steadily improved in design because the work of typists and word processors is a measurable (and increasingly expensive) commodity. ("Word processor" refers both to the machine and the operator, as "typewriter" used to.) The work of executives is measurable in the long run, but not in a way that can be clearly traced to the tools they use.

Action Office® system invented by Robert Propst.

Desks, chairs and typewriters as individual objects are not the problem. The problem is the office itself, designed on the basis of everything imaginable except an analysis of what it is for and how it is used. Few designers, and few heads of corporations, have seen any urgent need for such analysis. Office work is an abstraction. There is no reliable way to assess the effectiveness of the office, except in regard to its peripheral functions. Designers have established criteria for cold and warm offices, power-base offices and shirtsleeves offices, but no criteria for office performance in supporting work.

Parkinson was right: office work *does* expand to fill the time and space allotted to it. Moreover, it demands increases in both. The research that led to Herman Miller's Action Office® system was an effort to approach the problem from the standpoint of office system design.

The end of World War II brought with it such strong resistance to "easy answers" that answers of any kind fell instantly into disrepute. It became fashionable for college professors to begin courses by insisting that they didn't have all the answers. Some took pride in not having *any* answers. What was important, everyone said, was to ask the right questions. Indeed it is, but it can be frustrating to stop there. Robert Propst, the researcher and inventor who developed Action Office®, deviated from fashion in assuming that the old questions were answerable and that the problems they revealed were solvable.

He began with a battery of questions that is continually being enlarged and modified. How do people work in offices? Sitting or standing, or even lying down? Should a door be opened or closed and for how long? Should there even *be* a door? Where are phones best located? How often do exectives nap in offices? Should the practice be discouraged? How much office equipment is purchased and installed for actual work purposes and how much for purposes of conferring status? Is neatness necessarily an asset? Is it more efficient to converse with colleagues in your office or theirs?

The point was not that such questions, or others like them, had never been asked before. The point was that they had never before been regarded as answerable by design.

Similar concerns had been raised by management consultants, time planners and even, in the thirties, by efficiency experts, who imposed their answers on workers as if the workers were the equipment. Of course office behavior had been examined in literature by authors as different from each other as Arnold Bennet, Sinclair Lewis, Stephen Leacock, and Elmer Rice. But Propst was posing his questions with the idea that they could lead to designing and making some answers.

What they led to was a system of office components, panel hung or wall hung, that can be arranged in various configurations to meet particular needs and can be swiftly rearranged as those needs change. Similar questions about health care led Propst to develop a comparable materials system for hospitals, and to expand the office system into a system of factory work stations.

Because office work consists largely of processing paper, further questions arose, having to do with how information is displayed, recorded and passed on. Also with how it is hidden in drawers or under stacks of other information, often of lower priority. Propst calculated that a pile of papers higher than three inches is too high for productivity, and designed tambour roll-top work surfaces to inhibit such piling.

Although we are becoming a nation of "knowledge workers," there is still relatively little knowledge about the knowledge work environment. Because offices are such unproductive environments, we turn to alternatives as often as possible. A clinical psychologist tells me that, on the basis of statistics he has seen, sex can be defined as what people do before marriage. On the basis of statistics I am perfectly willing to make up, work can be defined as what people do out of the office. "He's at home today—he had a lot of work to do" is not an uncommon explanation of someone's absence from the office. Much of our work gets done in the Extended Office of restaurants, trains, airplanes, homes, phone booths.

Once I was researching a report on prisons at the same time I was working on a project for a manufacturer of electrical appliances, and I flew directly from a maximum security state penitentiary to my client's industrial research and development center. As I walked the corridors between laboratories and engineers' offices, I actually forgot where I was and wondered why there were no guards around. Too much work and not enough sleep, I thought. But when I returned the next morning, I realized it was neither fatigue nor jet lag that had confused me but the absence of major interior design distinctions between the prison and the equally inhumane research center. Both were characterized by a nearly total sensory deprivation. (There were some differences: the research center cafeteria, unlike the prison one, did supply table knives, and the engineers, unlike the prisoners, were allowed to go home at night.)

The Jewish *shatkhn*, or matchmaker, is the classic situation designer. Part of a solid and very clear tradition, the matchmaker is an excellent example of the design process at work. Her constraints were money, age, looks, social position. Within them, she provided a service with results more easily measured than those of any advertising agency or corporate graphics firm. She knew her clients and their needs and used the situation in which her clients found themselves as a basis for designing situations in which *they wished* to find themselves.

Although the individual marital matchmaker has largely vanished from American society, other kinds of matchmaking persist. Computer dating services, churches and synagogues, executive search organizations. . . .Both big government and big business have urgent and constant need for people whose only professional skill is in bringing other people together.

An especially rich document in the early literature of situation design is the U.S. Army's standing orders for Rogers' Rangers. Issued in 1759 by Major Robert Rogers, there are 19 orders, of which the first is "Don't forget nothing," and the last is "Let the enemy come till he's almost close enough to touch. Then let him have it and jump out and finish him up with your hatchet." Compare that, for clarity, directness, and anticipatory verve, with the last corporate policy memo you've read.

Some situations can be redesigned only through objects and vice versa. Time was when men wore jackets in hot weather not because they were embarrassed to be found in shirtsleeves, but because a jacket was the only repository of such appurtenances as eyeglasses, comb, keys, checkbook, appointment book, pens, address book, business cards and photographs of children.

Clearly what every man needed was a purse, just as every woman needed interior jacket pockets. Both were forbidden fashions. The solution to male purse envy came initially not from designers (who are often the last to perceive that there is a problem) but from photographers, who discovered that a camera case would hold, in addition to film and extra lenses, such items as dental floss and what airlines call "smoking materials." Soon camera buffs began loading their cases with personal property not manufactured, or even dreamed of, by Kodak. Many men took up photography solely in order to have a place to carry small purchases without

Shoulder attaché case designed by Bill Blass.

having to overdress. Finally someone discovered that he could buy a camera case without buying an instrument to carry in it. As long as nobody looked inside—and nobody ever did—the user could carry accessories on his shoulder, and ride with them in his lap, without having his masculinity questioned.

In the late 1960s Bill Blass designed for the Wings Company a canvas "shoulder attaché." It held legal pads, manila folders, books, and had an outside pocket for a newspaper or magazine. The first men to use these were, predictably, greeted with sarcasm ("Don't forget your purse, Mike."), followed by a guarded acceptance ("I can see where that might come in handy at times."). Final acceptance came when totebag-bearing men were stopped in the streets and asked bluntly where to buy them and for how much.

The Blass bag was sold mostly in department stores and cost around $30. Totebags today are sold almost anywhere except banks (where they are given away) and they cost from $30 to more than

$1,000. You can get them in vinyl, several grades of leather, canvas, fur, and even wood. The variety is dazzling, although I have seen none that surpasses in utility Blass' original, which unfortunately has been unavailable for years.

Once a verb, *tote* has gone on to become a noun, meaning a bag with which toting is done. Totebags cut across all levels of American society. Clerks carry their lunch in them. The dapper and macho president of a West Coast advertising agency has for years carried the firm's most important papers to and from meetings in a Danish school bag which he buys every few years in New York at a shop for children. The chairman-of-the-board of a large furniture company in the Midwest appeared at a board of directors meeting last winter carrying a khaki totebag from Eddie Bauer, a serious sporting goods store. Game bags sold by Abercrombie & Fitch and L. L. Bean have been pressed into executive service. An artist in New York never leaves his loft without his musette, a piece of army gear that turns out to be a totebag in a plain olive drab wrapper. Hunting World, a New York safari boutique, has for years advertised a jungle-derived totebag for people who have no intention of going on safari or even reading Hemingway.

Has the proliferation of totebags demolished the men's jacket industry? As a matter of fact, it has not. For, as the totebag phenomenon has grown, so has the paraphernalia men are obliged to carry. Standard equipment has been enlarged to include pocket calculators, pocket dictating machines, answering-machine beepers, computerized foreign-language vocabulary displays and Swiss Army pocketknives. As a result, men still need jackets, because without them there is no place to carry such appurtenances as eyeglasses, comb, keys, checkbook, appointment book, pens, address book, business cards and photographs of children.

An ingenious example of the product-situation cycle could be found in a Quebec waterfront hotel called L'Hotel Louis XIV, lamentably destroyed by fire a few years ago. At the Louis XIV, the term "private bath" meant what it means in many European hotels: the bath is yours but not yours alone, for it is also the private bath of the guest on the other side of the bathroom. This creates a

Third-floor bathroom in the Hotel Louis XIV sketched by Milton Glaser.

problem. If the bathroom has no inside locks, you have no privacy. But if the doors can be locked from the inside, one forgetful guest can lock the other out indefinitely and almost surely will.

Well, there were no locks inside the bathrooms of the Louis XIV, but tied to each doorknob was a three-and-a-half-foot length of leather thong to which a hook was attached. When you were in the bathroom you simply linked the two hooks together, holding both doors shut. There was no way to get back into your own room without at the same time unlocking the door for the other guest. It was memorable as the total integration of object and circumstance.

As a rule, bathrooms in themselves are notoriously ill designed, a situation that has been documented in great detail by a number of studies, most notably a massive one done at Cornell University in 1966. Bathroom fixtures may be our best illustration of the phenomenon Elizabeth Gundrey (page 110) talks about —the tendency for inadequate designs to spawn product lines to make up for their inadequacy. Bathroom boutiques and the bath departments of large department stores are all filled with conveniences that have had to be invented because the room itself was not properly designed in the first place.

Probably the worst designed feature of the bathroom is the standard toilet. As product design, it may promote disease by failing to permit complete elimination of fecal matter from the lower bowel. As situation design it is not even comfortable for reading.

The public lavatory presents special problems, most of them having to do with maintenance. Stores, restaurants, gas stations, transportation terminals—to say nothing of the planes and trains themselves—all provide public necessities that the management is unable, and sometimes unwilling, to keep clean. The pious injunction ordering all employees to wash their hands before leaving has connotations that make the whole experience less appetizing rather than reassuring.

Dirt is one problem of the public lavatory; vandalism is another. While this sometimes takes the form of stolen paper and dispensers ripped from the wall, it is more likely to be expressed as graffiti. The United States Forest Service has funded the development of improved outhouses, with both interior and exterior surfaces that resist writing and carving. In an unusually creative response to graffiti, a New York City public school at one point included in a maintenance man's duties the transformation of the word FUCK into BOOK, with a magic marker.

While the public lavatory may be seen as a situation desperately in need of design, it is frequently also an instrument of situation design. Gas stations discourage, or even prohibit, the use of restrooms by motorists who do not buy gas. The restroom thus becomes a sort of merchandising device, just as service used to be. The withholding of relief is often incorporated into the design of service station policy, snobbish stores, and even municipal governments. Enraged and offended by the presence of hippies (i.e., young people with long hair and blue jeans) the Aspen, Colorado, City Council blocked the construction of public restroom facilities, apparently in the belief that if the kids had no access to public toilets, they would go on to the next town, or even the next state. Neither nature nor rebellion works quite this way. The kids demonstrated, in more ways than one, and eventually a public facility was built, although I could never find it.

More complex and less charming than the Louis XIV bathroom is the Apollo space program. The design of a system for putting men on the moon and bringing them back could not have succeeded apart from the design of the situation in which those men would eat, shave, defecate, collect data and perform unprecedented acts in an alien environment. Nor could it have been brought off without the design of the situation on the ground: legions of men in white shirts who could be depended on to perform as reliably as the legions of machines they tended.

As spectacular as the product design is, the situation design in the space program is what has held the public imagination. "If we can get to the moon, why can't we abolish poverty?" has been scoffed at as a simplistic question, but it is not. The traditional offering of government is bread and circuses. With the success of the space program people began asking why the intelligence and skills of circus management couldn't be applied to the distribution of bread.

It is perhaps easier to perceive designed situations in circuses than in social service. Theatricality has never been limited to formal theatre, and designers need to take this fact into account. A contemporary kitchen is not merely a space where cooking is done; more and more it has become a space where cooking is *seen* to be done. The cook rises to the occasion not merely by the preparation of a consummate soufflé, but by the display of tools with which to prepare it.

The "family room" of the suburbs is similarly more (and in recent years less) than a place for the family to do things together. It is a spatial advertisement for a way of life, the builder's explicit provision for the consumers' tacit announcement of the lifestyle they have moved to the suburbs to acquire. During those occasions, rare indeed in most families, when people are actually clustered in the family room, they look very much like the families in the TV situation comedies they are watching there. (An English architect visiting New York once remarked that the most striking characteristic of Americans was that they looked so much like American advertisements.)

In the spring of 1964 the world's imagination was captured and terrified by the murder of Kitty Genovese in Kew Gardens, New York. Gratuitous murder in itself is not all that interesting. What was interesting enough for acres of analysis in print was the fact that 38 people had watched the murder from their windows without feeling impelled to do anything about it, not even call the police. This was instantly attributed to the callousness of the urban environment, the fear of getting involved. A friend of mine explained it another way: since the murder was played out in the street, viewers at their windows saw it as a stage presentation. They would no more intervene than they would rush up on stage to keep Othello from murdering Desdemona. If my friend was correct, the theatricality of that event came at precisely the time when theatre was becoming more lifelike and more highly participatory. In other words, as theatre took on the chaos and meaninglessness of real life, real events were approached with the detachment of an audience.

Another sort of case in point is the hotel architecture of Morris Lapidus. Often cited for the vulgarity of his designs, Lapidus is able to talk about them without being unduly defensive, and did so in an interview with John Margolies. More than most architects, Lapidus knows what he is doing and whom he is doing it for. About the clientele for the Fontainebleau Hotel, he asked rhetorically, "What do they want? What are they looking for? They don't want leisure. . . . They don't want homelike surroundings; they don't want the world's realities. They're looking for illusions. . . ."

Lapidus then addressed himself to the question of where the desired illusions come from. Not schools, museums, or European travel, he reasoned, but from the movies.

> This is a play . . . so I put them on stage at all points . . . to get into the dining room you walk up three steps, you open a pair of doors and you walk out on a platform, and then you walk down three steps. Now the dining room is at exactly the same level as my lobby, but as they walk up

> they reach the platform. I've got soft lights lighting this thing up, the captain steps up on the platform and, before they're seated they are on stage as if they had been cast for the parts. Everybody's looking at them; they are looking at everybody else. And I've been waiting for someone to say, "Hey, Lapidus, will you tell me why we have to walk up three steps, walk out on a platform and walk down three steps?" and no one has ever remotely questioned, Is this the way to do it? Now that's sheer nonsense. Why couldn't they walk right in? That's not what they want. They want to be on stage . . .

One needn't go to a hotel dining room. Theatre—or in any case theatrics—could be found at the family dinner table back in the days when families ate dinner together. Today, participatory theatre is more often found in designed selling situations. A San Francisco interior design firm announces "an old adage in retail design: people go to stores for sheer visual drama as much as they go to buy merchandise." I never had heard the adage, haven't found anyone else who has, and don't believe it anyway. There *is* a visual drama in contemporary retail merchandising (New York's department and specialty store display windows are among the most striking design details in the city) but what drew people to stores and still does was probably drama of another sort: stores were where things happened.

A store or showroom is less an inventory of products than a selling situation. It is dynamic, not static. People show each other things, talk, question, argue. This *is* theatre, and any set designer knows that the closer he works to theatre the less he needs to do anything "theatrical."

If a design is important in an immediate, rather than a seminal, sense—that is, if its importance is the direct effect it has on people rather than the effect it has on designs to come—then it is impossible to consider it apart from the situation. Large urban planning projects always fall into this category. The potential for great good or great harm is intrinsic to them, and one of those potentials is much easier to achieve than the other. Jane Jacobs' crusades, both

Selling situations today tend to be designed centrally (by package designers, advertising people, rack jobbers, etc.) rather than locally. But the proprietor of this 1924 butcher shop, authoritatively positioned behind the meat counter, did all the selling himself.

in print and in political meetings, on behalf of neighborhoods have always acknowledged design in situational terms. Her gift is to be able to see the whole picture, even though her notoriety has rested on the passionate conviction with which she could illuminate details. Seeing the whole picture, and knowing where the bodies you don't see are buried, is exemplified in her opposition to Westway, the billion dollar highway that threatens to strip Manhattan of much of its charm, many of its amenities, and the hope of rescuing its mass transit system. Identifying Westway as part of a 1929 plan by the New York Regional Plan committee, Jacobs observes, "Pieces of this plan. . . keep surfacing every few years. Nobody would ever consent to the insanity of doing the whole thing, and yet piece by piece it gets done." Designers have enough trouble just trying to control *things.* Situations are often uncontrollable and to make them otherwise suggests manipulation.

Store-window design (called "window dressing") used to consist mainly of changing the clothes on manikins four times a year. In recent years the vitality and imagination of window designers has enlivened city streets and reflected social changes: manikins are no longer all white or even all pretty.

When I discussed situation design with a marketing vice president, he recognized it as part of his job and said: "Managers are situation designers because they determine the vehicle, the setting in which communication takes place. This need not be manipulative—in fact it can't be. Manipulation occurs when you don't design. I mean a salesman can manipulate, but not a manager." The idea of designing situations professionally had a special appeal when guerilla theatre and other techniques of the politics of confrontation made manipulation a necessary, even attractive, evil to students (including design students) and to young people generally (including young designers). Advocacy planning, an inevitable response to civil rage, led the way to a sort of advocacy design. Some designers began to act as if they had constituencies instead of clients.

This was never a "movement," but a variety of more or less

Translucent fabric tent roofs let natural light into Bullock's department store in San Mateo, Cal. Designed by L. Gene Zellmer Associates; Geiger Berger Associates, structural engineers.

complementary activities. In England, Archigram was pushing technology into fantasy, creating with its plug-in and clip-on cities a kind of three-dimensional science fiction about the built environment. In the Arizona desert, Paolo Soleri's Arcology, rooted in high seriousness and religiosity, sought to exploit technology as an instrument of salvation. Organizations with names like rock groups, like California's "Ant Farm," became the design wing of social protest. In Philadelphia GEE! was developing educational materials on ecology and on "the public use of public land." Landscape architect Lawrence Halprin, who had earlier used choreography as a model for understanding urban spaces, produced a "scoring" technique for designing spaces in terms of the activities that go on in them, and vice versa.

But the *principle* of situation design has been with us from the beginning. The Garden of Eden was charming as landscape architecture but significant only as process. As set forth in the Book of

Genesis, the compelling design details do not lie in the vegetation, elegant though it was, but in the situation: a man, a woman, could have anything in the world they wanted, except the fruit. Under those circumstances there was only one thing in the world to want. Everyone knows that. If they tasted the fruit they were going to be in big trouble. But without the perpetual prospect of big trouble, there would be no eternal narrative to unfold. Every creator knows that.

However ancient a concept, situation design remains a fuzzy one. The Italian industrial designer Ettore Sottsass speaks of "the design of possibilities," a phrase I like better. Situations are cold and hard and once removed from dreams. Possibilities are nicer, and designing them is a lovely, daring prospect. To design *for* possibilities is a girl washing her hair, then waiting at home for the phone to ring. To design *against* possibilities is buying life insurance.

Both are dull ways of spending time.

But to design possibilities themselves is to open up new experiences. Sottsass referred to the Algerian government's interest in "the design of alternate possible Algerias." Certainly alternate possible Americas, families, ways of life and careers are familiar concepts to us, but they are rarely *design* concepts. Yet design is intrinsic to the issue: many of our most nobly motivated movements have foundered on design failures. In their zeal for an alternate lifestyle, hundreds of thousands have failed to distinguish between a lifestyle and a styled life.

Design is an abstraction. It is only when you unwrap it that you find the load of concretions it represents. Marianne Moore called poetry an imaginary garden with real toads in it. Design is a real garden with imaginary toads in it; the designer's role is to realize the toads.

It may seem odd to talk about the design of possibilities when we are so demonstrably inept at the design of probabilities. Our public streets are the discredited laboratories of design for pedestrian concern. Although in some states the pedestrian has the right of way, that is the only right he has. Everything is designed for the motorist. No, not for the motorist, for the car. Actually, streets are not that well designed for any of the three end users—pedestrian, motorist, or car. "Street furniture" is usually terrible, perhaps because cars don't have to sit on it. There are exceptions all over the country—in Portland, Oregon; in Minneapolis; even in New York. But for the most part, our benches, litter baskets, street lights, phone booths, and signs betoken a massive indifference to what things look like and how they work.

Architects used to speak of "making statements," but those statements were architectural or at least sculptural. The classic design statements have in an astonishingly literal way been sermons in stone. Apart from the function served (the Seagram Building, after all, contains office workers as well as design dogma), the buildings were preachy. But their text was narrowly limited to design, and the sins they railed against were the venial sins of eye and drafting table, and the virtues they exemplified were technologically sophisticated versions of the stonecutter's virtues.

The socially rooted design of possibilities is something else. It stands in the same relationship to classic design statements as protest demonstrations and love-ins stand to traditional Protestant sermons. Ant Farm was the illogical extension of Gropius's Lab Workshop at Dessau; design is no longer a cause but a tool for reaching the unreachable. Small wonder that when radical theologians were declaring that God was dead, radical designers were declaring that so was architecture. Alternate lifestyles, to the communal designers, mean alternate architectures, alternate *goods.* Utopia is mapped by the *Whole Earth Catalog,* which begins with a battle cry for the design of possibilities: "We *are* as gods and might as well get good at it." (An improvement on the more common, but unstated, premise: We are as goods . . . .)

It is tempting to see our cities as problems and to look for a grand plan that will solve them, but the rules of the game seem to preclude any such solutions. If instead we were to use design as a way of concentrating on making cities livable, they might begin to seem more like cities and less like problems. In any case, we would understand and like them better.

Situation designed by the New York City Transit Authority.

Take mass transit for example. Probably no single system had as much going for it, in a design, engineering, and economic sense, as San Francisco's BART, which had behind it the technology of the aerospace industry, a $790 million bond issue, a platoon of reputable designers, and a cluster of worthwhile motives. Its backers were motivated to solve the environmental crises, prevent the "Los Angelezation" of San Francisco, make job locations accessible to carless minority members, and serve as a model of how military technology could be pressed into humane service.

What went wrong? Everything. The answers are not in yet and may never be. But, apart from the technical problems that have kept the trains from running or from running on time, BART seems to have been superbly designed to serve the needs of some people who live in some other place in some other way.

People who *do* live in some other place in some other way can be found in New York City, waiting for buses. When, after a long wait, one finally comes, it is (1) jammed, and (2) followed closely by three empty buses. No other bus will come until the next pack of four.

A city bus driver, writing in *The New York Times* under an assumed name, explains that these situations are just as frustrating for drivers as for passengers. If he were driving bus number three, he points out, "the best step for me to take from the point of view of service would be simply to pull out of line, run down ten or fifteen blocks and begin picking up passengers there." Why doesn't he do it? Because Transit Authority rules do not permit him to pass waiting passengers at a bus stop or to run ahead of schedule.

The design of the buses themselves is notoriously inadequate, which is blamed in part on inadequate budgets (although designers must have something to do with it). But the system is as badly designed as the vehicles, and there is no budgetary reason for that.

Transportation is loaded with difficult situations, most of them having to do with human contact. Sealed from communication with others, the driver on a Los Angeles freeway invents ways to signal: waving, honking, displaying bumper stickers for Proposition 13 long after it has become law, holding his telephone high enough to show the adjacent driver that he has one. The driving itself—the quality of movement from lane to lane—is as much message as medium.

But public transportation forces an uneasy human contact that the user has to adjust to with very little help from designers. The experience of riding in a subway or elevator calls to mind Bertrand Russell's remark that much of modern anxiety stems from the time we spend in unnatural proximity to strangers without the preliminary sniffing that is instinctive in animals, including us.

Codes appear quickly in the absence of a workable design. The New York subway code is simple in theory but very difficult for strangers to learn. It says simply that no passenger may intentionally catch the eye of another passenger; if it happens inadvertently the offender must turn his gaze away instantly, careful not to let it land on still another passenger. What makes this difficult is that most subway cars are designed with the seats facing each other. So passengers have developed various strategies for dealing with the situation. Hiding behind a newspaper works well if there is enough room to open the paper. (Skinny tabloids are best.) Some passengers close their eyes to feign sleep, but the Transit Authority discourages this as an invitation to muggings. In the service of advertisers designers have located the ad space overhead, thus encouraging passengers to rest their gaze reverentially on pictures of cigarettes, hair oil and hemorrhoid medications.

The problem is even more intense in elevators. An elevator ride is an exercise in keeping one's distance without having any distance to keep it in. Here, passengers are permitted, almost required, to look at each other upon entering; but only for an instant. The elevator journey proceeds on the premise that passengers will all face the same way, like lemmings or television viewers. When that contract is violated, the effect is unnerving. If you want to see grown men cry, try facing the rear of an elevator and looking into the eyes of your co-travelers. Few people do that, but it has become common for riders to assume ambivalent positions along the side walls, facing right or left, exposing their profiles and the nerves of other passengers.

Escalators present still another set of choices. An elevator is a contained environment, like a ship. You can jog in it if you wish but it won't get you to your floor any faster. An escalator, however, operates simultaneously in two modes. Literally a moving stairway, it lets a rider take matters in his own feet, so to speak, and accelerate his speed by walking while riding. But he can do this only if the passengers ahead leave clearance. Some, standing to a different drummer, do not.

Our most advanced essential artifacts make us vulnerable to situation design failure. The telephone creates more problems than the elevator, giving A the means of reaching B, but giving B no chance to avoid that call without avoiding other calls he may want. Answering machines can take care of that one, but they give rise to more, and sometimes greater, difficulties.

"All we really have is time," writes Joseph Heller. "What we don't have is what to do with it." Because the design of possibilities is the design of what to do with time, it must combine vision with urgency. We have had visionary designers aplenty, but they generally fall into one of two categories. There are designers whose vi-

Coming and going in the Pan Am Building, New York City.

sions challenge everything we now know, but are based on a technology yet to be invented or a way of life few are tempted to pursue. At the other extreme are designers whose published work is visionary in the sense that it may never be built, not because it couldn't be but because it is more interesting on paper.

An example of a situational visionary (those who do not know jargon are condemned to invent it) is the Israeli-born architect Moshe Safdie, who came to public attention with Habitat, a housing experiment designed for Expo '67 in Montreal. Ada Louise Huxtable, writing in *The New York Times* the day the fair opened, said of Habitat:

> Just about every housing and building rule, precedent, practice, custom, and convention is broken by Habitat. This includes design, engineering, construction, trade union operation and the way people live. There have been two results. One was snowballing costs and technical problems. The other is a significant and stunning exercise in experimental housing . . .

Habitat in Montreal, designed by Moshe Safdie.

What was most important and striking about the Habitat concept was the astonishing variety growing out of technological standardization. Standardization connotes uniformity and the sacrifice of privacy. Safdie's vision was to see that contemporary prefabrication techniques were not only compatible with old world charm, they were the means of achieving it at large scale. The staggered units use each other as Mediterranean communities use cliffs. Each of them was conceived as a totally prefabricated concrete unit manufactured and fitted with prefabricated fiberglass bathrooms and an energy core, on site, then lifted into place. The project was to be massive, low cost, and assembled with the ease we have come to expect of mass manufacturing techniques. Instead it was fairly small, very expensive, and was bogged down in a process combining bureaucracy (federal, provincial, and Expo), handcraft, and show business scheduling. Nevertheless, it works and it tells us things.

There is a Frank Harris short story called "The Irony of Chance" about a scientist who gives lecture-demonstrations in which he makes a metal ball move by commanding it to. Because his power to do this is not always reliable, he hires a small boy to get inside the ball and make it roll.

One day the boy asks for the evening off, but the lecturer refuses, for he needs him. During that evening's performance, as the ball rolls across the stage, a heckler shouts from the audience that the scientist is a fake, that there is a boy inside the ball.

The lecturer denies it at first, then confesses, arguing vainly that the theory is valid nevertheless.

The incensed crowd reviles the scientist and demands a refund and a signed confession of fraud. Afterward, as the lecturer sits alone and dejected, the stage door opens and the boy enters: he had taken the evening off without permission.

The experience of Habitat is precisely that experience. It was not the product it was intended to be, but it was a compelling object lesson in what that product can be, and in the situation required to bring it about.

Probably each successful design teaches us something about the relationship between its components. This is no less true of situation design. When, during World War II, relocation camps were set up on the west coast of the United States, the lives of the Japanese Americans interned in them were disrupted both physically and psychologically. Although entire families were taken intact from their homes and farms, family stability did not always survive the trip. In the camps, parents and children ate in a common dining room and were governed by common rules set by the camp authority. Older children, then, were not dependent upon their parents for anything; rather both they and their parents were dependent on the same outside source. This had the effect of diminishing parental authority to the point where families quite simply came apart because there was no objective basis of need for them.

The design of objects, by its very nature, limits the use to which they can be put. Yet users are always extending the designer's intention: a fishing tackle box makes a perfect actor's make-up kit. The user in that case is an uninvited redesigner. But the conscious design of possibilities invites, indeed demands, the collaboration of the user as a sort of co-designer, a much less patronizing category than "consumer." And the designer also profits; the design of a possible tool is more rewarding than the design of a probable tool. The design of possible anythings is liberating, for it necessarily includes the possibility of being wrong. The physicist Niels Bohr had a horseshoe hanging above his office door. A visitor, noticing it, incredulously protested, "Surely you, one of the world's most distinguished scientists, don't believe that horseshoe will bring you good luck!"

"Of course I don't believe it will bring me good luck!" Bohr replied indignantly. "But I have been reliably informed that it will bring me good luck whether I believe in it or not."

"America was promises," said Archibald MacLeish. But promises can be broken. To the designers of the nation, America was possibilities. The Declaration of Independence carefully eschewed the promise of happiness, providing only for its pursuit. The framers, ahead of their time in design as in everything else, were concerned with process rather than with end product. Their designs were truly anticipatory, at once describing possibilities and protecting them.

If America is to keep the promises made of it in the eighteenth century, it is necessary to drastically redesign our domestic situation. No one seriously questions that. Economic experts cannot agree on how to design the nation out of an inflation and a depression simultaneously, but they all agree that it must be done.

Curiously, this is where we came in. Industrial design in America emerged in part as an answer to the question, "How do we get the economy moving again?" The answer appeared to be products that were new, or that looked new, or that were better, or that looked better. It wasn't such a bad answer then, but it isn't a very good one now. What needs to be designed now is the dynamics of the system, not just the shiny objects that roll off the end of it.

Are designers equal to the task? Of course they aren't; but they could be. Professional designers are almost always better than the products they design. Younger designers in particular wish to do something more socially useful than flooding the world with more, or even better, washing machines. Situation design is the most promising avenue to the kind of significance that many designers aspire to and many others claim to have attained.

Designers certainly *could* attain it. They are almost invariably more knowledgeable, more intellectually aware, and more decently motivated than one would expect them to be on the basis of design school curricula or design office employment criteria. (Indeed, these very qualities, sometimes combined with inadequate skills, often *disqualify* young designers from jobs with design offices or in corporations.) The history of the design professions is largely a history of overqualification, of men and women who have insisted on doing more than either clients or public ever asked for.

Yet how can even the most gifted designers contribute to solving problems of situation?

Edward Banfield observes that "social problems are at bottom political . . . and, except in trivial instances, are difficulties to be coped with (ignored, got around, put up with, exorcised by the arts of rhetoric, etc.) rather than puzzles to be solved. In coping with difficulties, formal analysis may sometimes be helpful, but it is not always so. (Would anyone have maintained that in the Convention of 1787 the Founders would have reached a better result with the assistance of a staff of model builders?)"

Designers are trained in the spatial organization of matter. Is that any reason to suppose they can deal with the human situations that defy Formica, curtainwall construction, universal joints, and even the typeface Helvetica?

Well, some of them can. Designers are hardly the chosen people, but they are qualified at least by default: everyone else is even worse equipped for designing situations than designers are. Social scientists know more, but have difficulty in making things work. Politicians get things done, but the things they get done are shaped by their desire to stay in office. Artists have more daring social visions, but it is hard to check out their validity in time to act. The designer is in respect to the world a paradigm of the human being in respect to civilization. That is, human beings are less well suited to any particular environment than most of the other animals that live in it. But as a result of belonging no place, we can live almost anyplace, including environments where other animals cannot live. Because designers are not especially trained for any field in particular, they can operate in a great many fields, and this adaptability is crucial.

Albert Szent-Gyorgyi says that molting is the principal means for mankind as a whole to adapt to radical new situations. He claims that people, like snakes, can grow only by bursting their outgrown skin—in our case, the outgrown skin of antiquated thinking and institutions. This implies risk of a kind inherent in design.

In a paper delivered to a conference on university planning, Harold Enarson, then president of Ohio State University, attacked the naive application of new technologies to management decision making. Pointing out that most planning research consists of efforts to "enlarge the data base," Enarson observed that "our problem is not the shortage of facts, but our general inability to grasp the significance of the facts . . . and our reluctance to do what needs to be done. . . . All too often [our] new tools and techniques create the illusion of planning and thus distract us from facing issues."

Richard Latham stated the problem years before: "People think they're planning when they're only plotting." Enarson identifies two models of planners: the Cook's Tour model and the Lewis and Clark model. The Cook's Tour uses planning as a way of maintaining order, of avoiding the unexpected, of keeping on schedule. The Lewis and Clark model uses planning as a means of directing the details of daily adventure.

"The Cook's Tour provides the illusion of planning in a world of imagined stability," Enarson says. "The Lewis and Clark tour is an adventure into the unknown."

Real design is always such an adventure. To plan in the Cook's Tour mode is to design by the numbers, which is not to design at all.

One of my college professors was fond of telling his students that the post World War II era was unique in that no serious thinkers were optimists. He must not have known about Bucky Fuller. Fuller believes that our problem is not the scarcity of world resources but the designed misuse of them. Dr. Jonas Salk once asked Fuller what he did. He said he was "engaged in what you might call comprehensive anticipatory design science."

"That's a description of my work, too," Salk replied.

When the protagonist of Kingsley Amis' *Lucky Jim* questions his fitness for a job he is told, "It's not that you have the qualifications for this job. You haven't. You don't have the disqualifications, though, and that's more important." Designers don't have the disqualifications of other professionals.

They have some disqualifications of their own, however. How can the same bumblers who make business cards too large to fit business card cases solve our problems? How can the people who design our detergent packages help us reduce visual pollution, to say nothing of more hazardous pollution?

They can help because the process they so imperfectly practice keeps forcing us to project and model what we wish to do, thus externalizing a universal human habit. Philip Morrison suggests that this may be the distinguishing feature that defines human nature. "We are beings who construct—singly and collectively—internal models of all that happens, of all we see, find, feel, guess, and conjecture about our experience in the world."

By way of illustration Morrison describes a species of termites that build an intricate architecture supported by a series of true arches. No matter how impressive this achievement, Morrison says, "there is never an arch present until one appears by chance; whereas when we build arches, or anything else, the arch is in some sense present before it ever exists."

Designers are professionally expert both in building arches and in realizing their presence beforehand.

Still the question persists: why should people who happen to be good at sketching, handling materials, creating physical forms, anticipating and exploiting new markets and new technologies, have any particular contribution to make to human situations? I suspect it is chiefly because design is a problem-solving process that begins with a human being.

That human being is not the constituent of the politician, or the consumer aimed at by marketing men, or the subject studied by social scientists, or the enemy of the general, or the character of the playwright, or the plaintiff or defendant of the lawyer, or the football coach's opposing coach. For a designer, the human being is himself or herself! The assertion is heretical, for it suggests "self expression," which is contemptuously dismissed as irresponsible. Nothing could be further from the truth. In design, the awareness of self is the beginning of responsibility.

But only the beginning. The next step is to take into account the needs of other people. For this purpose, the kind of research that designers can do, and have always done, has come back into fashion. The 1973 Nobel Prize in Medicine was awarded to Lorenz, Tinbergen, and Von Frisch, all of whom take an ethological approach to medicine. That is, they watch how an organism behaves in relation to its environment. Designers have always done this, if only because they didn't know how to do anything else. The kind of research that consists of looking around and putting yourself in someone else's shoes is not as superficial as it may sound. Putting yourself in someone else's shoes implies knowing a lot about yourself (and a fair amount about shoes too, for that matter).

The design of possibilities is initiatory rather than reactionary, a departure from the mode of operation in which the client shapes the problem because it is his problem. There are no existing clients for possibilities, at least no clients who already know they want them. Designers normally pretend to work like private eyes, waiting for the client to stagger through the door with a new case. It is romantic to view the consultant designer as a kind of Sleeping Beauty, doomed to inactivity until some corporate Prince Charming arrives with a wake-up kiss; but design isn't like that anymore and never was. The beauty of a design solution can never lie in wait; it requires active involvement all the way: initiating design and taking responsibility for it in all the messy situations of distribution and use that can never be recorded in the slide presentations that designers show to clients and to each other.

Surely it is necessary to redesign the situation of design itself. In the meantime there is no reason designers cannot initiate projects of their own and find sponsorship for them if necessary. A number of designers already do this. Benjamin Thompson, the architect responsible for snatching Boston's Quincy Market from the jaws first of defeat, then of literal restoration by the book, did not act as his own developer, but he did the next best thing. He found the developer, James Rouse. The project was architectural, but the design depends more on buying, selling, meeting, eating, playing and listening to music than on a series of harmonious facades.

Throughout this volume I have claimed that design solves problems. It often does. But when we call designers problem solvers, the connotations are very grand. An infusion of humility is useful. It helps if we remember that, to a person hungry for scrambled eggs, a short-order cook is a problem solver. It also helps to remember philosopher Abraham Kaplan's observation that we talk about problems when we mean predicaments, which cannot be solved and must be faced, lived through, or otherwise dealt with. One reason for believing that designers could professionally address social issues is that their primary competence lies not in the technicalities of a craft but in the mastery of a process that can help us solve problems or deal with predicaments. (One of our problems is how to design our own predicaments.)

That process consists generally of seizing on a purpose; defining the situation or problem; identifying constraints and organizing materials, people and events in a way that can be modeled and visualized in advance. More important than the sequence of steps, however, is the rooting of the process in needs and aspirations that are at once personal and public. Some of those needs and aspirations are downright prosaic, as are some of the designs that help satisfy them. But we ought not to disparage the means just because it can achieve ends that are not lofty. The same process that shapes our useful objects—cameras, buildings, furniture, bicycles, knives and forks—can be a tool for shaping how we live with them and with each other.

Making Connections: The Designer as Universal Joint

"What do you think of Eames?"
"I don't even know what they are." **Anonymous**

Mark Twain's novel *The American Claimant* begins with the arresting announcement that "no weather will be found in this book." It would be technically inaccurate to say that no designers will be found in *this* book, but certainly none has been discussed at any length up to this point. That might seem like a curious omission in a book about design, but our concern is more with the design process than with individual design professionals.

Paradoxically, *because* of that emphasis I want to examine the work of a particular designer, Charles Eames, who died in 1978. I choose Eames not because he is typical (which he is not) but because of the extent to which his career passed through and assimilated the stages of industrial design itself: product to environment to communication to situation. Other designers have gone part of the way with equal distinction, and young designers will carry design further than Eames did. But as a kind of working laboratory for the design profession, the Eames office is an unusually good source of insight into the design process at work.

Chances are that you associate Charles Eames with a chair, which is like associating Isaac Asimov with a book. Eames designed scores of different chairs (and many variations of most of them), but when people speak of "the Eames chair," they have one specific chair in mind, although not necessarily the same one. They mean the molded plywood "potato chip chair," the fiberglass side chair, or the leather lounge chair with ottoman. Eames' chairs and seating are so well known—from their appearance in magazines, museum collections, homes, offices, airports and prestige advertising—as to obscure the fact that he was never primarily a furniture designer. The furniture embodied a working philosophy as much at home in a chair as it was in a master plan for an aquarium, a film about symmetry, a toy, or a proposal for addressing some of India's economic problems.

As I indicated, I do not believe that any chair, however elegant, contributes significantly to life or solves problems that can be considered major by anyone who is not minor. However, the Eames *approach* to chairs (and to anything else) is an approach we can bring to activities more important than taking the weight off one's feet. Like most of us, he was not much concerned with chairs as such, but with connections, and with learning how to make better ones. In the narration he wrote and delivered himself for a film he and his wife Ray made to explain a storage system, Charles says: "The details are not details. They make the product. The connections, the connections, the connections."

If the design process is to become a force for making things right, we have to learn more about connections, for making them is intrinsic to every design from a submarine to the door hooks in the Hotel Louis XIV. Connections between what? In Eames' case, between such disparate materials as wood and steel, between such seemingly alien disciplines as physics and painting, between

Molded plywood chair designed by Charles Eames.

clowns and mathematical concepts, between people—architects and mathematicians and poets and philosophers and corporate executives. (Also between Charles and Ray Eames, who were husband and wife and full collaborators.)

There were no other designers quite like Eames, and still aren't. His uniqueness was not his talent, which was prodigious; or his intellect, which at times went almost unnoticed in favor of the elegance and wit that showed up in the work; or even the work itself, which materially affected how we sit, store things, build, play, communicate, teach and learn. The uniqueness lay in the design practice that the Office of Charles and Ray Eames maintained for more than thirty years. Eames was a practicing designer in the sense that others are practicing physicians; and such professionalism, as we have seen, is a commonly stated aim among designers. He was also a practicing designer in the sense in which others are practicing Christians, and that is uncommon and a more useful model for nondesigners. Plying his philosophy as if it were a trade, Eames practiced what many designers, and many of us fellow travelers, preach. That is, he attacked only problems that genuinely interested him, in the conviction that the solutions, if valid, would interest a great many other people.

This was not self expression, which Eames consistently disparaged as a motive in design and art, but a sense of the level at which meaningful connections could be made. "We like to think we're pretty special," he explained. "And, OK, in some respects maybe we are. And we like to think that we design for ourselves. And we do, we really do. But in the *important* ways we are really very much like a lot of other people. And if you are going to design for yourself, then you have to make sure you design *deeply* for yourself, because otherwise you are just designing for your eccentricities and that can never be satisfying to anyone else."

If the designer's self was sufficiently acknowledged and realized, it didn't have to be consciously expressed. But the process *was* expressed. Part of the craft and art of Charles and Ray Eames was to solve problems without obliterating all traces of them. Each solution articulated the need it addressed, so that the problem became part of the pleasure of its own solving, in the way that Robert Frost says a poem, like a piece of ice, should ride on its own melting.

This meant not only making connections but celebrating them. The problem of the molded plywood chairs was understood in a way that made separate parts, and therefore joining, possible. So the Eameses developed a way of exploiting high-frequency electronic bonding techniques to use rubber shock mounts as joints connecting the plywood seat and back to steel legs. Thus, both the problem and the solution are visually recorded in the chair's form.

The separately molded pieces were a compromise after years of painstaking, heartbreaking, and sometimes very funny attempts to produce a one-piece laminated chair. That compromise resulted in the "petals" for seat and back that are so important to the chair's success and, by now, to the vocabulary of design. A recurring theme in Eames' work and thinking is the creative acceptance of constraints, the satisfaction in pushing a material or an idea or a budget as far as it will go.

A related theme in the Eameses' work is simply work itself, hard work. Admiring designers today speak of "Eames quality," and wonder how it was achieved. It was easy, if you didn't mind its being almost ridiculously hard. *Quality meant that the details were not boring.* Because Eames and the architect Eero Saarinen had launched their design careers by winning a furniture competition, Charles once was asked about the trick of winning competitions. He admitted there was a trick. "It's really Eero's trick," he said, "but I'm going to break a rule and reveal it."

> This is the trick, I give it to you, you can use it. We looked at the program and divided it into the essential elements, which turned out to be thirty odd. And we proceeded methodically to make one hundred studies of each element. At the end of the hundred studies we tried to get the solution for that element that suited the thing best, and then set that up as a standard below which we would not fall in the final scheme. Then we proceeded to break down all logical combinations of these elements, trying to not erode the quality that we had gained in the best of the hundred single elements; and then we took those elements and began to search for the logical combinations of combinations, and several of such stages before we even began to consider a plan. And at that point, when we felt we'd gone far enough to consider a plan, worked out study after study and on into the other aspects of the detail and the presentation.
>
> It went on, it was sort of a brutal thing, and at the end of this period, it was a two-stage competition and sure enough we were in the second stage. Now you have to start; what do you do? We reorganized all elements, but this time, with a little bit more experience, chose the elements in a different way (still had about 26, 28 or 30) and

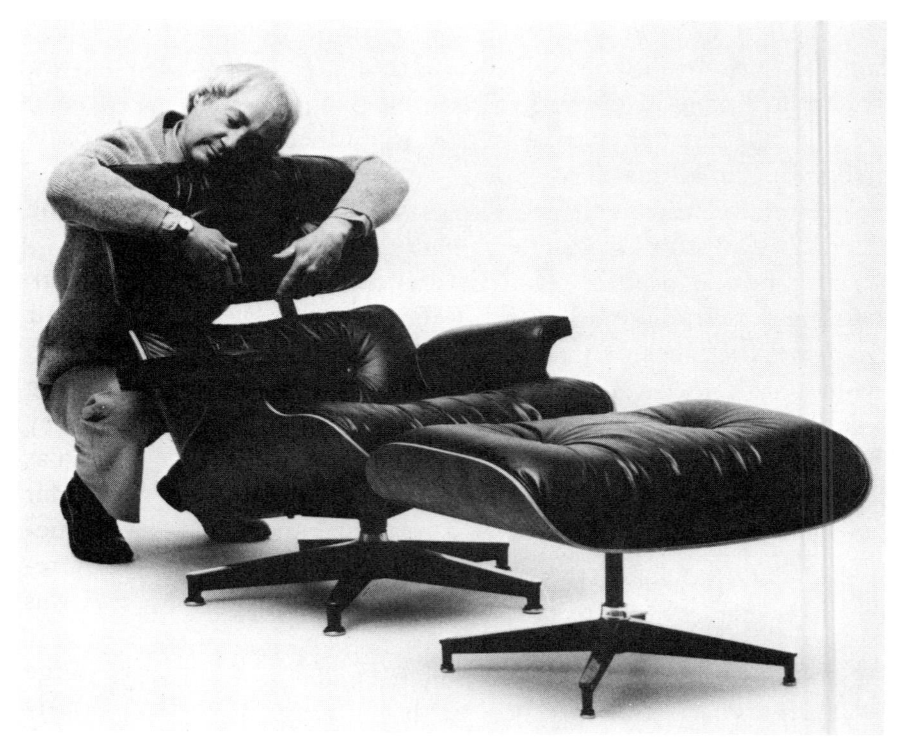

Lounge chair and ottoman designed by Charles Eames.

> proceeded: we made 100 studies of every element; we took every logical group of elements and studied those together in a way that would not fall below the standard that we had set. And went right on down the procedure. And at the end of that time, before the second competition drawings went in, we really wept, it looked so idiotically simple we thought we'd sort of blown the whole bit. And won the competition. This is the secret and you can apply it.

All that for a chair! And a chair sometimes described as "playful." In fact Eames was so deeply devoted to play that he couldn't distinguish it from work, although he believed passionately that "it is almost impossible to reconcile self-expression with the creative act." His preoccupation with constraints ran counter to the very idea of a "me generation." A tight and painful discipline brought him to grips with the most prosaic and minute problems, and if the final product sometimes had the appearance of indulgence, it was always an earned indulgence. However, Eames did not take on any truly mass-market products, which would have entailed prosaic problems of a lower order of prose. He seemed instinctively to eschew constraints that militated against excellence while embracing those constraints that were ultimately liberating.

We can see that process at work if we look at where the work got done. Design offices, their walls colorfully postered and their desks productively cluttered are—because of the mysterious and visually rich stuff designers work with—almost always interesting to nondesigners. Clients look for excuses to hang around the way Broadway angels like to hang around theatres: it is as close to fun as many of them come in their working lives. The Eames office, far from being an exception, was an exaggeration of this phenomenon. People remember their first visit there as vividly as they remember what they were doing when they heard that Kennedy was shot.

The building in Venice, California, was a surprise: a large white old fleet garage situated unceremoniously and anonymously along some abandoned railroad tracks. You entered a reception room dominated by an Indian tapestry, some Eames furniture and an IBM electric typewriter (there had been a battered Remington manual until an IBM client noticed it and quickly sent a replacement). The reception room and a small accountant's office were as close to business atmosphere as the place got; the rest was something else.

From the reception room you were led directly into a confrontation with toys varying in scale from tiny tops (dozens of them) to a glass case through which you could cause thousands of polyethylene balls to drop through a configuration of hundreds of aluminum pins, producing a musical demonstration of distribution curves. Hard-bitten corporate visitors could not suppress their delight, but were uneasily aware that they were in a place where the bottom line was not a line at all, but more like an enveloping cloud, and where the bottom was deeper than they were used to.

Office of Charles and Ray Eames, Venice, California.

What they felt, I think, was the tension between complementary opposites: charm and hard labor, play and work, delight and deadlines. The paradox was intrinsic to the operation. For all the people working here—there might be about twenty; for all the distractions—there were about a thousand; for all the variety of things to see and touch and play with, there was an inescapably unifying element: *Charles was in creative control of whatever was going on.* And what was going on—as in many design offices—was an object lesson in how to sustain creative vision in group activity.

It was also an object lesson in the design emphases we have

been looking at in this volume: objects (chairs and speakers and storage units), environments (showroom interiors, a model of an aquarium and a tankful of fish), communications (massive exhibitions, films in progress, tens of thousands of slides) and situations (a plan for enriching the art curriculum at MIT, a scheme to provide entertainment for people waiting in long exhibit lines).

A design office is not really a place, it is a collection of designers. The Eames office was a diverse, dedicated group propelled by an inordinate energy upon which inordinate demands were made. Designers speak of working *en charette,* a French phrase meaning "in a cart." The usage is derived from Beaux Arts architecture students who finished their drawings in the mornings while riding to class in carts. It means working night and day until a project is done right. For the Eames office *en charette* was a way of life. On any given day, and night, it would go something like this.

In a fully equipped shop at one end of the building prototype furniture is being made; only a few feet away, in a temporary film set, three scientists from the Rand Corporation (located a couple of miles down the road) are listening attentively to directions from Charles. Today they are actors, but they are the same scientists Eames regularly turns to for advice and criticism. They have previously taped a discussion of the intricate system by which the heart supplies blood to the lungs. The designers rewrote the transcript and are now restaging the scene, on camera, hoping to keep in the prepared script the flavor of the spontaneous talk.

Elsewhere in the office a social scientist is reporting his estimate of crowd handling for an exhibition, while a marine biologist checks out the octopus in the aquarium.

The simultaneity of design activities is exhilarating. In the same shop and at the same time people are drafting, sawing, sewing, welding, sketching, photographing, editing film, reading, writing, making presentations and above all talking—talking to clients, physicists, composers, materials experts, writers, mathematicians, production foremen, each other.

Detailed models and mockups are being built, graphic material is collected and sorted, scripts are written and storyboards prepared at the same time, furniture is built, admired, tested and rejected, steel pipes are welded together, ribbon is cut.

At the center of all this there is always a model, for modeling is both metaphorically and physically central to the process. It is how designers project in three dimensions. Models can be superficial, just as products can. Some design models are cute, idealized miniaturizations. Although their purpose is to show the client, who presumably cannot read plans, what the final project will look like, what they usually show instead is what the project would look like only if people behave the way architects wish they would. Models of this kind are not appreciably different from those that children make with kits. Frequently they are treated as esthetic objects. Some architectural offices have "model rooms" to which clients are brought for visiting, as if the model were an invalid aunt. Some models go directly from the model shop to the client's reception area, where they are displayed as a three-dimensional artist's rendering of things to come.

Anyone engaged in a creative act risks failure, but there is no law against trying to avoid it. That is one use for models in design and in everything else: they are working tools for anticipation. Those in the Eames office were also like scientific models, instruments of discovery. By manipulating them you could not only figure out what would work and what would not; you could learn new things.

Eames had at any given time only a handful of clients, who assumed that whatever he did would be something they wanted, even if it was not exactly what they expected. When Westinghouse set out to get him to do a film on their design program, Eames proposed instead a film establishing the breadth of the corporation's products. The result was *Westinghouse ABC*, a visual and musical introduction to the company as producer of a staggering range of objects and systems, of which consumer appliances were a relatively small part. The film became an extremely effective communications tool for Westinghouse, but they never did get the one they originally asked for.

For Herman Miller, manufacturer of Eames chairs, a specialized

Fabric stretched over cylinder in Eames Aluminum Group® chair.

and affluent market made difficult connections possible. In a line of chairs called the Aluminum Group®, the seat pad's two outer layers of fabric and an inner layer of plastic foam are combined through electronic welding. The entire seat pad is stretched across a two-sided die-cast aluminum frame that is cylindrical at top and bottom. The ends of the seat pad are turned up over the cylinders in each corner and held by tension. Supported by metal only at the corners and sides, the fabric seat is a slung bolt of softness juxtaposed against the elegant hardness of the frame. Both qualities are visible and tactile; end and means are equally discernible and almost indistinguishable from each other.

This approach to furniture uses advanced technology to achieve lightness, as semiconductors do. The lightness in turn makes mass-produced objects personal not by "personalizing" them but by incorporating free individual use into their design. As the English critic Peter Smithson wrote, "Eames chairs are the first chairs which can be put into any position in an empty room. They look as if they had alighted there. . . .The chairs belong to the occupants, not to the building."

This mobility did not preclude an appropriate richness: Eames chairs are not all comfortable for long, but they are all sensuous. "Less is more," Mies van der Rohe said (although who remembers where and when he said it?), and the phrase became a rallying cry for modernist starkness. Ornament, which had been banished by somber Bauhaus decree, had its excommunication confirmed by epigram. "Less is a bore," replied Robert Venturi some twenty-five years later, and the Post-Modernists set out to prove it by rummaging in history for old ornament to put on things. Unfortunately, but predictably, they also proved that *more,* in the right hands, is at least as boring as less, and gets there just as fast. The pleasure in Eames chairs, however, was not derived from a philosophical stance dictating the simple addition or subtraction of ornament, but from a sense of enjoyment to be shared.

Enjoyment, however, was not simple hedonism. In a talk given in London in 1963, Charles Eames described his attitude toward design freedoms and restraints, deploring the apparent tendency of design schools to orient their teaching toward the former at the expense of helping students develop an awareness of the latter. This is much more than the insistence that "you have to learn to walk before you can run," for it recognizes discipline not merely as preparation but as a quality intrinsic to the work. Design is itself an act of faith and discipline; where the restrictive lines are not clearly delineated, the designer must find them or draw them.

> For an example of how limitation works in the creative process, consider a sculptor attacking granite with hand tools. Granite resists such attack violently: it is a hard material, so hard that it is difficult to do something bad in it. It is not easy to do something good, but is extremely difficult to do something bad.
>
> Plastilene, though, is a different matter. In this spineless material it is extraordinarily easy to do something bad—one can do any imaginable variety of bad without half trying. The material itself puts up no resistance, and whatever discipline there is the artist himself must be strong enough to provide.

> I feel about plastilene much as the ancient Aztecs felt about hard liquor. They had the drinks. But intoxication in anyone under fifty was punishable by death, for they felt that only with age and maturity had a man earned the right to let his spirit go free, to self expression. Plastilene and the airbrush should be reserved for artists over fifty.

First-hand learning is probably an unbeatable basis for discipline; Eames was self-taught in an almost preposterously literal sense. "My father was a lot older than I," he liked to say and after waiting for the amused shock to dissipate he explained, "He was about sixty when I was born." As a boy Charles discovered some of his father's antique photographic gear and instruction booklets stored in an attic, and was overwhelmed by the possibilities. He figured out how to use the materials, became a photographer of sorts, and reported that "I had been making wet plate pictures, mixing my own emulsions, for more than a year before I found out that film had been invented!"

In design, as in all our lives, deadlines are a constant constraint. When the Eames office designed a speaker (not successful in the marketplace) under a crushing deadline, Charles said, "What we did here isn't any big deal. It was a case of the best we could do between now and Tuesday." Then, to wipe out any vestige of implicit apology, he added: "But the best you can do between now and Tuesday is still a kind of best you can do." But any kind of "best you can do" is rendered difficult by the loss of traditional restraints, Eames explained.

> The tradition, or lore of an art, helps the artisan bridge those gaps where he lacks sufficient current information upon which to base a decision. Innovation which takes place outside the restraints and assets of tradition often suffers from the lack of small, obscure, but vital bits of seemingly unrelated information which accumulate slowly as a tradition develops.
>
> For example, a plastic cup seems like a very reasonable thing. Who would have guessed that one would actually miss feeling the heat of the coffee or the coldness of the

> lemonade? Or that the constant neutral temperature of the material would give some of the disoriented feeling of novocaine in the lip? Who would have guessed that one would be disappointed in not hearing it clink when set down, and feel slightly cheated at the thought of its bouncing when dropped?

Charles and Ray Eames began making films early in their career and these have long been smash hits in the design underground, for there were not, and are not yet, many places to see them. The films and exhibitions gave the Eameses opportunities to make statements they found personally worth making, regardless of the client; but they never thought of themselves as filmmakers. To them, film was a tool, and although they pioneered some techniques we see on screen routinely today, the techniques were specific solutions to specific problems. For the CBS special "The Fabulous Fifties" Eames introduced the fast-cutting of still photographs as a way to cover a decade through existing photojournalism; for the United States Exhibition in Moscow, Eames used seven screens instead of one, on the premise that if the Russians saw enough simultaneous images of American lifestyles it would be harder for them to dismiss the film as fiction.

Multi-screen projection, like Plastilene, tempts a designer to indulgence, and Eames was aware of that danger in himself. He cited the law of the hammer: "If you give a small boy a hammer, he will discover that everything he encounters needs hammering. If you give an old man several screens. . . ."

An early film called *Bread* is a study of a large variety of baked dough. The camera pans lovingly over an appetizing assortment of bagels, rye, challah, stone-ground whole wheat. Even a cursory survey of the subject yields a variety of meanings. Bread is the staff of life. It is also the hip term for money and the symbol of Christ's body and of the Diaspora.

When I was a child my father used to like to take a slice of American white bread, roll it in the palms of his hands, and hurl it at the wall. It would bounce. If instead of hurling it he simply *pressed* it against the wall, it would stick like rubber cement. While that seems, out of context, like wildly aberrant behavior, it really was a rational, if amused, expression of contempt for a culture that regarded this product as edible.

Bread is a visual feast of alternatives to the stuff my father threw. One of the film's goals was to explore the educational uses of all of the senses. While the film's orientation is to objects, its commitment is equally to ideas—particularly to the idea that learning is a sensuous process and that teaching ignores that fact at its peril.

Most industrial and graphic design offices do exhibitions, and many of these deal with scientific, or at least highly technological, content. But they are almost invariably predicated on the notion of split expertise: the "content expert" has the knowledge and the "media expert" boxes it. Usually, then, when exhibition design deals with ideas, it deals with ideas for which the designer takes no initial responsibility. His role is not to decide what to say but to package it effectively, whatever it happens to be.

Eames exhibitions are different. No other design office I know of has worked on exhibitions as scientists would work in a laboratory (although perhaps the only scientists who actually have are Eames office collaborators). Discovery, not translation, was the business at hand, a business carried out by serious craftsmen, experimenting, formulating hypotheses, and testing. Small wonder that Jeremy Bernstein's *New Yorker* profile of Nobel laureate physicists Lee and Yang was recommended by Charles as required reading for "designers and anyone else interested in how the creative process really works."

The prevailing ground rules for exhibition design have to do with making things simple and attractive. While this is an aim in communication generally, it has a special importance in exhibition design, a craft in which the simple and attractive easily become the simplistic and slick. As an intellectual activity, exhibitions would seem to have modest possibilities. The audience is not concentrating alone with a book, or held captive by the darkness of a motion picture theatre, with only one place to look. The viewer is on foot and—in the case of fairs, for example—may have been on foot for hours. Neither the time nor the form seems appropriate for the explanation of tough concepts.

Yet explaining tough concepts is precisely what the Eames exhibitions have consistently set out to do.

When in 1946 the Museum of Modern Art held a show of the furniture designs of Charles and Ray Eames, the Eameses designed the exhibit itself (which unfortunately was not the case in the 1973 show of their work in the same museum). It included a mechanical demonstration of how energy was transferred from the seat of the chair to the leg, for they were already concerned with communicating process. How strongly the concern persisted was clear in 1961 with the opening of "Mathematica: A World of Numbers . . . and Beyond," an IBM-sponsored exhibition at the Museum of Science and Industry in Los Angeles.

"Mathematica" brought together a large number of features that had come to characterize the Eameses' work generally. It was loaded with detail, largely in the form of a massive "history wall" that related significant mathematical developments to each other and to other developments. It was participatory, not because viewers activated mechanical exhibits by pressing buttons, but because they performed operations that led to understanding. And it was *enjoyable.* The idea that science can be fun had been advanced soberly by textbook publishers in the forties and was reflected in the Armed Services training films done by Walt Disney and various of his imitators. In these the fun was additive, either blatantly in the form of irrelevant, and ultimately distracting, jokes, or subtly in the form of patronizing hypothetical problems. (There is a scene in a John Horne Burns novel in which a high school mathematics teacher assigns the dissection of a skirt as a classroom problem calculated to interest girls and their mothers.)

But Eames never set out to make science fun. He set out to help people experience the fun that is science. "Mathematica" may not really make mathematics easier, but it makes it clearer, inviting our perception of the elegance that mathematicians are always talking about.

The pleasure in the exhibition is like the pleasure mathematicians find in their subject, which explains why professional mathematicians and small children can share the experience and repeatedly do. The viewer of the "Mathematica" peep show film modules is not told about the concept of symmetry or even shown it, as with snowflake illustrations. Rather, the viewer is brought into confrontation with the concept's visual (and therefore literal) meaning and philosophical (and therefore mathematical) implications.

A key exhibit in "Mathematica" was the information wall, or history wall or histomap—a linear display of parallel developments over time that became a kind of Eames exhibition trademark, one that is not universally popular with other designers, many of whom complain that it is cluttered with excessive informational detail. It is excessive in the sense that it delivers more information than can be easily taken in by the average viewer. This is not "information overload," the simple effort to overtax the viewer's brain to the point where its capacity increases as a muscle does under isometric contraction. It is the presentation of enough data to reveal any datum in context—to make connections and reveal them so that he who runs may read.

Because of the depth of material, he who saunters may read at considerable length—more than most viewers will take time for; but so what? For the IBM pavilion at the 1964 New York World's Fair Eames created an area called "Scholar's Walk," deliberately designed to make use of rich material there wasn't room to include in the main exhibits. "Scholar's Walk" was directed to people whose interest had been so fired up by the exhibition that they were interested in finding out even more. Beside more hard information, it contained anecdotes, curiosities, cartoons, poems, etc. dealing with computers. It was not intended to be a crowd pleaser and wasn't.

Another function of the information walls is to let the viewer in on the process that has preceded what he sees. Just as a writer's instinct is to lay all his materials on the table, a designer's impulse is to spread them out on a wall. There may be another reason for such a busy graphic approach and for Eames' heavy use of histomaps. Apparently in Eames' view the untidiness of life could be dealt with best by tying it together in as many places as possible. Designers frequently prefer more order, even if it comes at the expense of substance. In Eames' world, however, there was quite simply a hell of a lot going on; and we come to terms with it not by trying to minimize the number of objects in the universe but by trying to establish relationships between them. (A less sympathetic explanation is that he could never make up his mind what to throw out.)

Formats like the history wall have caused Eames' exhibits to be called "wordy." Eames was both more visually perceptive and less hostile to the word than most designers. Eames office exhibits do not necessarily use more words, but they use them differently—as writers use them, rather than as design elements. They appear in the context of other words, photos, artifacts, and films. The words are to be read; if they are not, the exhibit is to that extent a failure.

That notion—that exhibits be read—was carried as far as anyone could wish in "A Computer Perspective," an exhibition dominated by a sixty-foot wall that was a sort of self-contained museum of computer pre-history. This may suggest a presentation so esoteric and highly specialized as to rule out of the audience everyone but data-processing pedants, and some of the material in that show was arcane indeed. But a lot of it—maybe most of it—was accessible to everyone, whether schoolchild or theoretical physicist, who has some curiosity about how things get the way they are.

The thing in this case was the computer. The exhibition traces, three-dimensionally, the events, people, ideas, machines, and problems that led to the development of the computer in 1950. In structure the exhibition is a combination of a tightly plotted suspense story and a Robbe-Grillet novel. A very clean line of development is fleshed out with hundreds of anecdotes, letters, parables, mathematical demonstrations, photographs, drawings, books, and artifacts ranging from mechanical monkeys to mechanical brains.

Three strains of development are plotted on the wall: logical automata, statistical handling devices, and automatic calculators. The story is that the computer as we know it (or more commonly don't know it) resulted from the merger of dancing dolls, census card sorters, and adding machines. This surprising hypothesis is made convincing by the amount, richness, and authority of the documentation. And the documentation was achieved through the same process prescribed for academic scholarship: original research.

There is another respect in which Eames exhibitions resemble conventional scholarship: the ubiquity of the footnote. Much of the information presented in these exhibitions takes the form of asides, elaborations, clarifications, illustrative anecdotes, and other material conventionally supplied in footnotes. Footnotes have been much maligned as the hallmark of pedantry, but actually they are often as not pockets of high reader interest. In the case of the Eames shows the interest runs so high that it seems to endanger the message. Won't some people get so caught up in the fascinating isolated fragments that they will miss the thesis? Unquestionably a good many will, but the highest level of participation surely consists in getting fascinated by the pieces and connecting them for oneself.

How does a designer get from the process of molding plywood to the process of tracking down scientific history? By making connections, by modeling the situation. When asked by the Massachusetts Institute of Technology to help infuse their technologically heavy curriculum with art, Eames rejected the idea of additional art courses or fine arts programs as "an esthetic vitamin concentrate." Instead he designed an alternative situation, a program for enriching the communicative capabilities of students to the point where they could experience the esthetic possibilities in *their own discipline.*

The situation he designed had two essential parts. The first called for each academic department to include a unit of teaching assistants whose first allegiance was to the departmental discipline, but who were also gifted and trained in film, graphics and writing. Their responsibility was to produce packets of current information that would keep everyone within the department aware of what was going on. The best of the packets would be made available outside the department, and the best of those would be distributed outside the university.

Work done by these units was supposed to "*arrive at* as well as *convey* insight," thus precluding the creation of still another campus media center to prepare slides on demand for instructors who wished to beef up nonvisual material.

The second part would involve each student; for each, near the end of his MIT career, would join one or two other students in teaching something about their major specialties to an elementary school class for a semester. The teaching could take the form of films, exhibits, lectures, games, models—whatever the team needed to make what they knew and understood meaningful to children. "If the MIT student is going to learn anything about art," Eames argued, "he will learn it here."

The entire design repudiated the conventional approaches to "art enrichment," which mainly consisted of three kinds of programs. One gives students massive doses of high art (no one gets a diploma without taking "appreciation" courses to guarantee that he has heard, if not listened to, Beethoven's Ninth Symphony and looked at, if not seen, a Dutch master or a reproduction of one). Another is an egalitarian attempt to "reach the student where he is" by running him through courses in rock and roll, horror movies, great graffiti of the sixties, etc. A third is the studio approach of encouraging the student to "do it himself" on the grounds that his "it" is as valid as anyone else's. (It may be valid but it can be unrewarding and terrible.) Eames' design, which called for appreciation of the esthetic character of the student's own discipline, included a favorite Eames idea: the university as a "found object," a collection of traditions and facilities already on hand that can be transformed by fresh perception.

Is that designing or thinking? Designing *is* thinking. The Library of Congress has received a $500,000 grant to acquire and make accessible Charles Eames' papers, drawings, original films and production materials. Eames wrote very little, but his ideas were refined through the years in occasional lectures supplemented by films, or vice versa. In 1970–71 he was appointed Charles Eliot

Norton Professor of Poetry at Harvard, which gave him an opportunity to distill his thoughts into six illustrated presentations, each loaded with nuggets of film and anecdote. Among the best of them, and central to his philosophy, was a consideration of the circus as an example of apparent license on the surface of a phenomenally tight discipline. (The Eameses were both devoted to circus arts and one of their finest films, *Clown Face,* was made as a training film for the Clown College in Sarasota, Florida.)

A circus is a nomadic society, Eames said, and the more colorful the nomadic society, the more rigid its mores are likely to be. The lot boss, he explained, used to drive around the lot in a sulky, dropping stakes according to the count of the horse's hoofbeats. The stakes marked where the quarter poles were to be placed. It was a ceremony that ended in the big tent.

In the *Bulletin of the American Academy of Arts and Sciences,* Charles wrote:

> The layout of the circus under canvas is more like the plan of the Acropolis than anything else; it is a beautiful organic arrangement established by the boss canvas man and the lot boss . . . . The concept of "appropriateness," this "how-it-should-be-ness," has equal value in the circus, in the making of a work of art, and in science.

A circus is an object lesson in what Eames advised us all to do: take pleasure seriously. The circus looks like self expression and is not; it pushes against limits; it takes its beauty from a disciplined mastery of details and from the connections between them. In those respects it is the classic designed situation.

Index